"Up-to-date, comprehensive, and highly relevant to scholars and students studying Arctic urbanism and governance. Urban and resource development projects in marginal communities under multiple pressures require new multi-actor and participatory governance that integrates international sustainable development mandates with local and Indigenous knowledge and perspectives while giving nature a voice. Cities and governments elsewhere should pay attention to what these Arctic cities learn."

Peter Hemmersam, *Professor of Urbanism, The Oslo School of Architecture and Design, Norway*

"This book serves as a powerful reminder that the challenges of transforming the urban Arctic are not as exotic or exceptional as they might seem. Instead, they closely mirror the struggles faced by other ordinary cities worldwide. The authors present an innovative perspective on multi-level governance for understanding local approaches to sustainability in the urban Arctic, offering valuable insights for planners and regional developers alike."

Agatino Rizzo, *Professor, Arctic Five Chair in Architecture and Planning at Lulea University of Technology, Sweden*

"Frequently contextualised at the margins of national peripheries or even as a resource frontier for European green transition, questions of sustainability loom large for many Arctic communities. With its focus on urban spaces in the European Arctic, this work provides insights into facilitating transitions from global policy to local implementation suggesting practical strategies for urban decision-makers tasked with implementing sustainability targets within their own unique challenges of climate and remoteness."

Corine Wood-Donnelly, *Associate Professor of International Relations and the High North, Nord University, Norway*

"Remote Arctic cities, like cities everywhere, face a difficult challenge in addressing the threat posed by climate change. This book examines the tools northern cities are using to cope with the crisis. The focus on participation and transnational cooperation distinguishes this work, ensuring that it is a valuable contribution to a growing literature."

Robert W. Orttung, *George Washington University, USA*

"Sustainable development promises that all good things can be combined, if we just do it right—globally and, hence, also in the Arctic. This volume takes the fragile conceptual marriage at its words: after examining the challenges facing remote urban centres (another oxymoron), the authors identify operational steps towards bridging the gap separating Arctic cities from global mainstream discourse."

Ulrik Pram Gad, *Senior Researcher at Danish Institute for International Studies, Denmark, Co-editor of Politics of Sustainability in the Arctic (w. J Strandsbjerg, Routledge, 2019)*

Sustainable Urban Development in the European Arctic

Focusing on cities in the European Arctic, this book consolidates research on sustainable development, local and urban governance, and transnational cooperation in the region. It examines to what extent there is transnational cooperation between urban areas in remote locations and how it can be enhanced to better align with global sustainable development policies to successfully implement goals such as the 2030 Agenda for Sustainable Development and the Paris Climate Agreement. Based on field research in seven cities in the European Arctic, Rovaniemi, Kolari, Nuuk, Akureyri, Tromsø, Kiruna, and Luleå, the authors explain why approaches to sustainable urban development differ between geographies, how policies relate to other local and global strategies, and to what degree the European Arctic is normative for remote regions at large. This book contributes to important conceptual debates on local governance and transnational cooperation by examining the benefits and potential issues of applying theoretical models of multi-actor engagement and participation in isolated populations. It argues that the participation of local actors in decision processes may encourage a better harmonisation of sustainable urban development approaches in the European Arctic and will have a greater impact at the global level if aligned transnationally. This book will be relevant to researchers, social scientists, policymakers, practitioners, and NGOs in the fields of global governance, sustainable development, sustainability research, and environmental studies.

Dorothea Wehrmann is a senior researcher at the German Institute of Development and Sustainability and has a background in sociology and political science. She obtained her PhD in 2017 from Bielefeld University with a thesis entitled "Polar Entanglements? (Critical) Geopolitics of the Changing Polar Regions in Inter-American Perspective" (Routledge, 2019). One of the key findings of her PhD research was that local perceptions have been little considered in dominant discourses on the Arctic and Antarctic and related territorial disputes, which inspired her to develop the SUDEA research proposal together with Arne Riedel and Michał Łuszczuk in 2018. Since the project's kick-off in November 2020, Dorothea has been principal investigator and co-lead of the SUDEA team together with Michał Łuszczuk. Constant throughout her career has been an interest in societal responses to global challenges. Her research interests and respective publications relate to cooperative frameworks in policy-making processes, legitimacy studies, and (critical) geopolitics.

Michał Łuszczuk is an Associate Professor at the Institute of Socio-Economic Geography and Spatial Management, Maria Curie-Skłodowska University. He holds a doctor habilitatus degree in political science and has been a co-leader of the SUDEA project. He is a member of the Committee on Polar Research, the Polish Academy of Sciences, serves as a delegate of Poland to the IASC Council, and was elected as a member of the IASSA Council (2017–2021). He has been a Fulbright fellow 2015–2016. His research interests include the geopolitics of polar regions, socio-economic adaptation to climate change in the Arctic, Polish polar policy, and the history of Polish polar research.

Katarzyna Radzik-Maruszak has a background in political science and law and is currently an Associate Professor at the Institute of Political Science and Administration at Maria Curie-Skłodowska University. She holds a doctor habilitatus degree in political science. She is a member of the Working Group on Participation and Civic Dialogue, operating under the Polish Minister for Civil Society. Katarzyna has experience in comparative qualitative research in European local governments. She was a visiting scholar at Tampere University in Finland from 2014 to 2015 and a research fellow at Charles University in Prague, Czech Republic, in 2021. She has been the head of several scientific projects. Her research interests include comparative public administration, local governance, and citizen participation.

Jacqueline Götze is a political scientist with a background in international relations. She works as a researcher at the German Institute of Development and Sustainability (IDOS). Jacqueline wrote her PhD thesis at the University of Bonn on the relations between Sámi organisations and the European Union (EU) while understanding these relations as an example of Indigenous peoples' participation in transnational policy-making. In her PhD thesis, she found that the participation of the Sámi as the only Indigenous peoples within the EU is only limitedly formalised, which influences the Sámi right to self-determination. Jacqueline's research interest focuses on the participation of marginalised groups, Indigenous peoples' rights, transnational cooperation, and green colonialism.

Arne Riedel is a lawyer and senior fellow at the Ecologic Institute, where he coordinates the Ecologic Legal team as well as the Institute's activities on Arctic issues. Within this topic, he focuses on Arctic environmental governance and has worked since 2012 on numerous Arctic policy projects for German ministries and agencies, as well as for the EU Commission and the WWF International Arctic Programme. Arne has been an ethics advisor to two Horizon 2020 projects and was also a member of the scientific advisory panel to the German MARE:N research programme "Coastal, Marine and Polar Research for Sustainability" in 2021. In his work on climate governance, since 2016, he has been a part of the German and EU delegations to the negotiations under UNFCCC, Kyoto Protocol, and Paris Agreement and, since 2020, a member of the Kyoto Protocol's Compliance Committee enforcement branch.

Routledge Studies in Sustainable Development

This series uniquely brings together original and cutting-edge research on sustainable development. The books in this series tackle difficult and important issues in sustainable development including: values and ethics; sustainability in higher education; climate compatible development; resilience; capitalism and de-growth; sustainable urban development; gender and participation; and well-being.

Drawing on a wide range of disciplines, the series promotes interdisciplinary research for an international readership. The series was recommended in the *Guardian*'s suggested reads on development and the environment.

Interdisciplinary Perspectives on Planetary Well-Being
Edited by Merja Elo, Jonne Hytönen, Sanna Karkulehto, Teea Kortetmäki, Janne S. Kotiaho, Mikael Puurtinen, and Miikka Salo

Good Education in a Fragile World
The Value of a Collaborative and Contextualised Approach to Sustainability in Higher Education
Edited by Alan Bainbridge and Nicola Kemp

Greening Higher Education in Europe
Institutional Transitions to Sustainable Development
Magdalena Popowska

Implementing Sustainable Cities
Edited by Sylvie Albert, Jeremy Millard, and Manish Pandey

Sustainable Urban Development in the European Arctic
Dorothea Wehrmann, Michał Łuszczuk, Katarzyna Radzik-Maruszak, Jacqueline Götze and Arne Riedel

For more information about this series, please visit: www.routledge.com/Routledge-Studies-in-Sustainable-Development/book-series/RSSD

Sustainable Urban Development in the European Arctic

Dorothea Wehrmann, Michał Łuszczuk, Katarzyna Radzik-Maruszak, Jacqueline Götze and Arne Riedel

Routledge
Taylor & Francis Group
LONDON AND NEW YORK

earthscan
from Routledge

First published 2025
by Routledge
4 Park Square, Milton Park, Abingdon, Oxon OX14 4RN

and by Routledge
605 Third Avenue, New York, NY 10158

Routledge is an imprint of the Taylor & Francis Group, an informa business

Funded by Deutsche Forschungsgemeinschaft (DFG, German Research Foundation) Project number: 426674468.

British Library Cataloguing-in-Publication Data
A catalogue record for this book is available from the British Library

ISBN: 9781032254593 (hbk)
ISBN: 9781032257549 (pbk)
ISBN: 9781003284864 (ebk)

DOI: 10.4324/9781003284864

Typeset in Times New Roman
by codeMantra

Contents

Figures

Tables

Preface

We are living through turbulent times, and the future looks increasingly bleak: at a global scale, multiple and interconnected crises are posing threats to human and environmental security in all its dimensions. Political structures and cooperation formats that have guided human coexistence on this planet since the Second World War are under increasing pressure. "The world is at crossroads" is a popular saying these days that also applies to the Arctic regions, which have been moving into an environmental disaster also prior to the crises that have materialised in the early 2020s.

In 2018, the authors of this book followed the aspiration to better understand how the global goals agreed upon in the 2030 Agenda for Sustainable Development and the Paris Agreement can be pursued in line with local realities. The main motivation behind the research project proposed back then was to learn from political approaches in the European Arctic on how concerted efforts across governance levels and regions can be advanced to address common challenges such as climate change. Since 2018, realities have changed significantly—also in the European Arctic. The aspiration to understand potentials and challenges for concerted efforts in remote regions, however, remains ever more important: "People want peace. [...] this is an international norm"[1] and "without concerted effort, there is a risk of a surge in major conflicts".[2]

This book summarises the results of our research conducted between November 2020 and August 2023 in the seven cities that the research team investigated in the European Arctic: Akureyri, Kiruna, Kolari, Luleå, Nuuk, Rovaniemi, and Tromsø. While the main research motivation did not change, the authors of this book had to adapt the research design to the impacts of the COVID-19 pandemic and the war against Ukraine. With the world at a crossroads, it is difficult to imagine how the broader context will shape future political developments in the European Arctic. This book, however, provides new and important insights into how concerted efforts can be advanced via (transnational) cooperation, strengthened local capacities, and inclusive participatory processes. Moreover, in this book, we propose a model to understanding why local approaches differ in the European Arctic, which, as is our hope, will be useful to better align global visions with local realities also in other remote regions.

Sustainable Urban Development in the European Arctic is a key reading for researchers focusing on the Arctic regions, from the field of sustainability research to urban studies. It opens new insights into political and institutional constraints imposed on cities, different types of available resources, and the relationship between cities and societies, including transnational forms of such relationships. The scale and dynamics of climate change in the Arctic make sustainable development a kind of imperative in adaptation and mitigation activities in the Arctic regions and in relation to the Arctic regions. This book further contributes to the conceptual debates on (transnational) cooperation and local governance by examining hindrances of and possibilities for applying theoretical models of multi-actor engagement across governance levels and participation in remote regions.

In addition to the scientific community, this book will also be of relevance to practitioners engaged in policy- and decision-making processes in countries with territory in the European Arctic, in the European Union, and in the Arctic Council. In particular, this book is relevant to practitioners with strategic responsibilities in the field of international relations (such as diplomats) or foreign aid (such as development professionals), as we identify gaps and inconsistencies in current legal frameworks at the regional, national, and local levels and outline potential adjustments, particularly in local governance approaches, to facilitate greater policy coherence.

Notes

1 Rt. Hon. Mark Pritchard MP, Vice-President and Special Representative on the Arctic and High North, OSCE Parliamentary Assembly, at the Arctic Circle Berlin Forum 2024.

2 Vision of Humanity. *2024 Global Peace Index*. Available from: https://www.visionofhumanity.org/maps/#/ [15 June 2024].

Acknowledgements

The research presented in this book was funded by the German Research Foundation (DFG) under the project number 426674468, and by the National Science Centre (NCN), Agreement UMO-2018/31/G/HS5/02448. The authors are very grateful for the independence provided by the DFG and NCN, which gave us the space and time to critically investigate our assumptions, to develop new ideas, and to conduct research in this project without any limitations to our thinking. We also greatly appreciate the trust and support of our host institutions: the German Institute of Development and Sustainability (IDOS), the Ecologic Institute, and the Maria-Curie-Skłodowska-University (UMCS). This book could not have been written without the engagement of our 80 interviewees who shared their time and insights with us. We are very grateful for their trust and helpfulness.

We are particularly grateful for the invigorating exchanges with our colleagues and friends, who inspired our thinking through challenging interventions at different points in time. We owe much gratitude to Mathias Albert (Bielefeld University), Steffen Bauer (IDOS), Stephan Klingebiel (IDOS), Robert Orttung (George Washington University), and Silke Weinlich (United Nations System Staff College) who reviewed our research proposal and supported the discussions of our research results in different research phases. We also want to express special thanks to our IDOS-colleagues Christine Hackenesch, Jonas Hein, Anna-Katharina Hornidge, Irit Ittner, and Michael Roll who shared very valuable comments on earlier drafts and presentations of our research. We are also very grateful for the management support provided by Mark Theisen and Christiane Weller and for the editing of some interview transcripts by Sophie Schlopsna and Jil Mast. We want to thank the communication and library teams at IDOS, particularly Conny Hornschild and Alex Fante who supported the publishing process of this book and Heike Großer for the compilation of the literature considered in Chapter 8. Many thanks also to Katharina Schaarschmidt, who kindly designed the graphics in this book, and to the reviewers from Editing Press for the careful language editing.

We are further indebted to the many insightful comments made by colleagues who read earlier drafts of the book's chapters: Nadezhda Filimonova (University of Lapland), Lena Gutheil (IDOS), Arto Haveri (Tampere University), Christoph Humrich (University of Groningen), Stefan Kirchner (University of Lapland), Hannes Hansen-Magnusson (Cardiff University), Robert Orttung (George

Washington University), Gregory Poelzer (Luleå University of Technology), Andreas Raspotnik (Fridtjof Nansen Institute), and Gary Wilson (University of Northern British Columbia).

Many thanks are also due to all of our colleagues who commented on our research at conferences, especially to Monica Tennberg (University of Lapland), Tobias Jakobi (Georg-August-Universität Göttingen), and Frank Sejersen (University of Copenhagen)—your questions and remarks very much supported the advancement of our preliminary ideas. We are also grateful for the in-depth discussions with our colleagues from the German Political Science Association (GPSA/Deutsche Vereinigung für Politikwissenschaft, DVPW) Thematic Group Polar and Ocean Politics.

This book illustrates the personal commitment and dedication of all colleagues involved. Over the course of four years, the authors of this book worked closely together to (re-)design the research, conduct fieldwork, analyse all data, and to draft, review, and edit all chapters—while working in different institutions, in different countries and passing through different stages in their lives.

Arne is immensely grateful for the support of his family, and, most of all, his partner Sophie, without whom the research and publication would not have been possible. The birth of their daughter Lilly Marie has given extra meaning to working towards a more sustainable and inclusive future, and she has brightened life beyond compare. Arne's father could not see the result of this work, but his dedication to fairness, his humour, and rich life remain an inspiration to live up to; he is sorely missed. Last but not least, Arne thanks his SUDEA colleagues who have been nothing short of wonderful with their hard work, dedication, and support (including by their families). Working together with them has allowed for the navigation of difficult times, made every meeting something to look forward to, and created a wonderful research family.

Dorothea owes a debt of gratitude to her family, in particular her husband, Sebastian. Work on this research and book project would not have been possible without his understanding, patience, and unlimited support. She also thanks her sons Aram and Emil for all the joy and energy they spread every day. She is also most grateful for her sister Astrid and her mother-in-law Roswitha—two incredible women who provided extra hands and arms whenever needed. Dorothea dedicates this book to her parents Christine and Werner who are deeply missed. This book would not exist without the extraordinary team spirit and commitment of all SUDEA colleagues, who collaborated with vigour and enthusiasm very often also beyond official working hours. Thank you!

Jacqueline wants to thank first and foremost her SUDEA team colleagues for the always kind, motivating, and supportive working environment. Special thanks go to the principal investigators, Dorothea and Michał, for their amazing leadership and to Arne and Katarzyna for their tireless commitment to the project and their humour—there has never been a meeting without a good laugh! She also wants to thank her partner and best friend Chris for always being there!

Kasia is extremely grateful to her SUDEA team colleagues for the kind, inspiring, family-friendly, and very supportive working environment. Special thanks go to the principal investigators, Dorothea and Michał, for their invitation to join the research team. She also thanks Jacqueline and Arne for their optimistic perspective on solving problems and their great sense of humour. She is also grateful to her husband, Grzegorz, and her sons, Tymon and Filip, for their support and patience.

Michał extends his heartfelt gratitude to his family—Monika, Bartek, Szymon, and Laura—for their unwavering support and patience. Additionally, he wishes to thank Prof. Wojciech Janicki for his invitation to the Institute of Geography, Socio-Economic, and Spatial Management of UMCS and for his invaluable support. Michał expresses his sincere appreciation to Małgorzata Południuk, Joanna Ciesielka, Maria Rymczuk, Anna Kiełb, and Agnieszka Ryś for their contributions and assistance during the project. And the last, but not least, Michał thanks the SUDEA team for the exceptional cooperation in exceptional times.

The authors declare that they have no known competing financial interests or personal relationships that could have appeared to influence the work reported in this book.

Abbreviations

2030 Agenda	2030 Agenda for Sustainable Development
AMF	Arctic Mayors' Forum
ECOSOC	The United Nations Economic and Social Council
EU	European Union
LGMA	Local governments and municipalities
LNOB	The principle to leave no one behind
MLG	Multi-level governance
NGO	Non-governmental organisation
Nordic Model	Nordic Model of governance
IR	International relations
UArctic	University of the Arctic network
PGS	Polycentric governance system
SDGs	Sustainable Development Goals
SUD	Sustainable Urban Development
TMNs	Transnational municipal networks
UN	United Nations
UNFCCC	The United Nations Framework Convention on Climate Change

1 Introduction

Why Sustainable Urban Development Matters in the European Arctic

1.1 Global Visions and Sustainable Development Approaches Are about to Fail

In 2015, the United Nations Member States agreed upon two global visions for the future: The Paris Agreement and the 2030 Agenda for Sustainable Development.[1] These two global governance instruments are based on a shared understanding of global challenges. The pursuit of the global goals agreed upon, however, is subject to cooperation and interpretation at national and local levels, which often do not connect well. This book takes this observation as a starting point to explore the fundamental question how societies observe, interpret, and approach places that are subject to substantial transformations.[1] We argue that if humans agree on the interpretation of shared challenges, this can become the basis for cooperation and collaboration.[2] However, different and changing interpretations shape policy-making on sustainable development in the European Arctic (Pram Gad *et al.* 2019). These different interpretations are determined by perception, knowledge, awareness, and attitude (Sudarmadi *et al.* 2001), and subject to change that is inspired by interactions with other humans and nature (Berghöfer *et al.* 2022, Tennberg and Strauss-Mazzullo 2023). The respective processes of observation and interpretation result in imaginaries that change over time and—similar to lenses—"allow one to refocus and defocus" (Buitrago *et al.* 2016, p. 6). These imaginaries thus guide the negotiation of priorities and policy-making on sustainable development that may result in competing, non-aligned visions for the future and respective actions.[3]

The 2030 Agenda and the Paris Agreement are global visions that are based on a shared imaginary of how the future of humanity at large should look. As normative frameworks, they guide actions of practitioners in the signatory countries and shape respective changes in societies. Ten years after the adoption of both agreements, it is clear that progress has been insufficient to meet the global goals. The Sustainable Development Goals (SDGs) experienced a reversal in progress (Kranholdt 2022) and it is unlikely that global warming will be limited to 1.5°C above pre-industrial levels (United Nations 2023). Even if the global visions agreed upon in 2015 are about to fail, it remains an immanent interest of humans to shape their future according to their current and future preferences. The challenges defined in the 2015-agreements are global in scope and require cooperation and

DOI: 10.4324/9781003284864-1

collaboration—even with "accelerated fragmentation between societal groups and between places" (Böhme *et al.* 2022, p.7). (Cooperative) Actions, however, will continue to be driven by interpretations and their change.

Both visions are anchored in the sustainability-paradigm. Sustainability builds on the assumption that human development cannot be achieved without sustaining the environment, which is known as the human-ecosystem-equilibrium (UNEP 2005). The concept of sustainable development merged discourses on environmental protection and development. It has its roots in the 1972 UN Conference on the Human Environment, which is often perceived as the start of global environmental policy. In comparison to sustainability, which is the paradigm that describes the long-term goal,[4] the concept of sustainable development has a process-character and features pathways to achieve sustainability. The advancement of the targets differs across policy fields and countries, monitoring frameworks and accountabilities (Haug and Taggart 2024).[5]

To achieve sustainability, the 2030 Agenda with its 17 SDGs features the pathway that all countries must act on the five interlinked Ps: people, planet, prosperity, peace, and partnership. Both the 2030 Agenda and the Paris Agreement follow a universal approach and do not differentiate between the geographical locations of countries (see Chapter 2). However, the principle to leave no one behind (LNOB) that both agreements emphasise acknowledges societal inequalities that result from "discriminatory laws, policies and social practices" (UN Sustainable Development Working Group 2024). In our context, a third international instrument, the New Urban Agenda (2016), which seeks to establish interlinkages between global agendas (UN-Habitat 2024), further recognises differences between urban centres and rural places. Thus, while all three visions are global in scope, they do not neglect our world's heterogeneity, power relations, and the different capacities and cultural, environmental, economic, and social priorities in societies. To achieve sustainability, the visions should guide actions at all governance levels and across regions, and for that, the goals need to be interpreted similarly at all governance levels to advance concrete actions and cooperation (SDG 17).

With the concept of sustainable development being widely referred to across policy fields, it can be seen as a "coalition magnet" (Béland and Cox 2015) that unites actors to strive for shared goals. However, the concept of sustainable development is also being used strategically in policy-making to drive priorities by and within countries (Thisted and Gremaud 2020). In this way, the concept has been criticised for its multiple meanings (Balestreri *et al.* 2023), as an "empty signifier" (Brown 2016) and as an umbrella term for conflicting sustainable development approaches (Wehrmann 2016b). With the global goals about to fail, it is notable that the selective translation of the "goals à la carte" (Reinar and Lundberg 2023) does not seem to help in overcoming the struggle between competing visions for the future (Laclau and Mouffe 1985 cited in Pram Gad *et al.* 2019). In contrast, from a geopolitical perspective, the concept of sustainable development and the goal to achieve sustainability has been reviewed as "another neo-colonial way for the West to dominate the rest by imposing standards limiting prospects for development" (Banerjee 2003, Sachs 1990 cited in Pram Gad *et al.* 2019), illustrated amongst others in debates on Green Colonialism[6] (Normann 2021). Competing visions thus range across governance levels, policy fields, countries, and regions. Visions of sustainable development, the contributions of the actors involved, and the outcome of the related negotiation processes are further affected by the inherent nature of policy-making as a political process, which is shaped by strategic interests and power dynamics (Brand *et al.* 2021). Accordingly, sustainable development is often investigated as both, a process and an outcome (Sachs *et al.* 2019, Simkiv *et al.* 2021).

In this book, we do not assess the impacts of societal approaches to urban development, the diverse understandings of sustainable urban development[7] nor sustainable development practices. Instead, in what follows, we will explore why societal approaches to urban development differ. More specifically, we investigate why policies for sustainable urban development are not aligned across governance levels in remote regions. This is the central question guiding this book.

One important lever to limit the intensification of the climate crisis is to develop urban spaces more "sustainably", as acknowledged in global instruments such as the 2030 Agenda for Sustainable Development (SDG 11: Sustainable Cities and Communities). Cities account for 70% of energy use and cause 75% of emissions worldwide (United Nations 2021). Particularly, regions that experience a strong urbanisation trend and are of unique importance in the global climate system are encouraged to build up city environments that match with the global goals adopted in 2015—such as the circumpolar Arctic. We focus specifically on urban places located in the European Arctic to better understand the challenges and potentials for pursuing the global visions agreed upon. Urban places (Brescia and Marshall 2016), and particularly those located in the Arctic, are examples of fast changing places that are subject to various challenges amongst others due to the rapid warming of the region (Rantanen *et al.* 2022). The circumpolar Arctic is moreover a remote region. Research on urban places in the Arctic regions has slowly grown in the past decade[8] (see, i.e., Aleksandrov and Dybtsyna 2024, Berman and Orttung 2020, Petrov *et al.* 2017) but Arctic cities are still little considered in urban studies and in the literature on remote regions. As we show in this book, however, the causes for the lack of progress in pursuing global visions such as "sustainable urban development" mirror and relate to the global and multi-level governance challenges identified also in other research areas. Bringing them together allows for a cross-fertilisation of knowledge and, as it is the hope of this book, contributes to a better understanding of how local approaches[9] evolve and how imaginaries and cooperation can become drivers of change that help to overcome the "existing ambition gaps" to pursue the global goals (Fuhr *et al.* 2018, p. 4).

1.2 Pursuing the Global Goals in Remote Regions

This book is driven by two assumptions: we expect that the remoteness of regions is an additional challenge for negotiations of different approaches to sustainable development to take place, often resulting in the domination of strategies developed in top-down processes. At the same time, we assume that in the circumpolar

Arctic[10] long established collaborative and often transnational[11] frameworks empower local governments to shape their futures and to implement participatory processes, which support the interpretation of the global goals at the national and local levels. However, as we show in this book, even in the European Arctic, participatory approaches are less advanced than we expected, and policy-making for sustainable development seems more often to be driven by top-down approaches.

Remote or peripheral regions,[12] in general, can be seen as examples for places that are often left behind: they are characterised by rurality, sparse population, and scarce infrastructure (Lindberg *et al.* 2020), often resulting in limited human and social capital and limited access to public services (education, health system). The related geographic and economic remoteness[13] often also result in structural remoteness in terms of political power (Berman and Orttung 2020). In the Arctic context, the harsh climate describes another factor for the region's remoteness. Environmental conditions hinder access to Arctic cities, connections between and also mobility within the Arctic regions in the North American Arctic, European Arctic and Russian Arctic, which greatly differ. When framing the Arctic as remote, the region can be characterised as externally and internally remote: externally, because of its distance to other regions, and internally, due to the remoteness between Arctic regions themselves as transportation routes often focus on North-South connections connecting the Arctic with the capitals and greater cities in the South (North-South axis, mirroring internal colonial structures).

Remote regions are still little considered in global governance research. However, since the 2010s, the circumpolar Arctic region is increasingly imagined as a "global Arctic" (Keil and Knecht 2017). This imaginary builds on the observation that climate change in the Arctic has global consequences and on the interpretation—and popular saying—that what happens in the Arctic does not stay in the Arctic and what happens outside the Arctic has significant effects on the Arctic regions (NATO Parliamentary Assembly 2017). In opposition to the concept of an "Arctic exceptionalism",[14] the circumpolar Arctic is thus increasingly seen as being globally embedded in environmental, economic, political, and social changes and as a "laboratory" and "barometer" for changes at the global scale (Bertelsen 2019, Khare and Khare 2021, Wehrmann 2016a).

Due to rapidly proceeding climate change, the Arctic landscapes and temperatures have significantly changed with dramatic effects on the environment that is home to more than four million—Indigenous and non-indigenous—people. Their lifestyles, (traditional) livelihoods and demography[15] are affected not only by these environmental transformations but also by the related economic and political changes. Access to valuable resources and growing tourism and shipping offer economic prosperity, but in the Arctic regions, economic development is intertwined with the environmental protection of a unique and globally important ecosystem (this normative trap is known as the Arctic Development Paradox, see Łuszczuk *et al.* 2022). Economic activities also affect the cultural traditions and housing of people in the Arctic, which have resulted in inner-state territorial conflicts.[16] There is a growing sentiment that human and environmental needs are subordinated to economic development and that land-use in the "Arctic periphery" is dominated by decisions made in the distant capitals of the countries bordering the Arctic Ocean and not in the Arctic regions themselves (McCauley *et al.* 2022). Political agendas, such as the Green Transition (Nystø Keskitalo and Götze 2023), further spur the sentiment of Green Colonialism in some regions (Normann 2021), as the Arctic regions are increasingly attracting industries that seek to develop renewable energy in the Arctic. Therefore, people in the Arctic increasingly feel that their needs are also subordinated to global interest to limit climate change, even though the climate crisis is caused by the consumption behaviour of people located elsewhere (Mathiesen 2023). As Hemmersam argues, however, "these concerns sustain a picture of the Arctic as a passive victim of climate change and potentially deemphasise the agency of local communities and governments in the pursuit of sustainable urban policies and planning" (Hemmersam 2021, p. 165).[17] Moreover, since February 2022, the Arctic regions are increasingly affected by geopolitical turmoil, with "no clear way forward" (Koivurova *et al.* 2022, p. 3). Due to Russia's full-scale invasion of Ukraine, circumpolar cooperation is experiencing a turn and established forms of cooperation are currently reordering (Methi and Wehrmann 2023, Wehrmann *et al.* 2022). Also political priorities have been reshuffling: security and military tensions have become great concerns in and across the Arctic regions (Wall and Wegge 2023).

Humans observe, interpret, and act accordingly—also in remote regions, of which the Arctic is an example. While the pursuit of global goals is of existential relevance for the Arctic, the re-interpretation of these is shaped by the region's global embeddedness and the manifold changes that the region is undergoing that are interpreted differently by the eight "Arctic Countries" with territories located above the Arctic Circle.[18] Further, the Arctic is not a homogeneous entity, but must be understood as "a region of regions" (Gamble and Shadian 2017, p. 143): environmental, economic, political, and social changes materialise differently in the European, North American, and Russian Arctic. It is thus no surprise that contestations are visible within and between imaginaries that shape Arctic visions (Steinberg *et al.* 2015) and also policies for sustainable urban development. However, considering the Arctic's global embeddedness and the rapid changes that the Arctic regions are undergoing, we argue that a better understanding of these competing visions and the related imaginaries is needed to advance a common understanding and cooperation across governance levels and regions.

As stated earlier, if humans agree on the interpretation of shared challenges, this can become the basis for cooperation and collaboration, which is needed for the pursuit of the global goals. In view of the Arctic, the concept of sustainability has been reviewed as "reaffirming the position of central actors", particularly of nation states and their concerns with economic development (Pram Gad and Strandsbjerg 2019, p. 250), whose "power operates through the construction of the social worlds in configurations of sustainability" (Sejersen 2019, p. 94). This observation stands in stark contrast to the multi-stakeholder and participatory approach envisioned by the global agreements to pursue the global goals (cf. Chapter 2). It also conflicts with the pluralistic stance to stakeholder engagement promoted by the New Urban Agenda that demands "capacity development as a multifaceted approach that addresses the ability of multiple stakeholders and institutions at all levels of governance" (United Nations 2017, p. 37). This demand builds on the understanding that the global goals need to be re-interpreted in participatory approaches at the national and local levels to result in aligned actions. Or said differently: if strategies to pursue the global goals are not negotiated among state and non-state "reactors" in the UN Member States, these strategies will be perceived as domination, will be contested and/or not be followed.

1.3 Pursuing the Global Goals in the European Arctic

This book focuses on the European Arctic to investigate why policies for sustainable urban development are not aligned across governance levels in remote regions. The more recent literature on sustainable urban development focuses mostly on the global trend of urbanisation, on the vulnerability of cities to climate change and on indicators for measuring the sustainable development of cities in the Arctic (Berman and Orttung 2020, Petrov *et al.* 2017). Beyond the Arctic, existing literature discusses the roots and dimensions of the sustainable development concept for urban areas (Wheeler 2022) and practical dilemmas (Metzger and Lindblad 2020). Brenner's book on New Urban Spaces and Hannigan and Richards's handbook on new urban studies offer new ideas on how to re-think and theorise urbanisation to understand urban transformations at the present (Brenner 2019, Hannigan and Richards 2017). There is a growing body of literature investigating why approaches to sustainable urban development differ based on empirical research and some already focus specifically on remote regions like the Arctic (Hemmersam 2021, Laruelle 2019). When aiming to align approaches across the Arctic, however, a better understanding of the different contexts in which Arctic urban development takes place is crucial. In this way, our book sheds light on political and institutional constraints imposed on remote cities and on the different types of available resources and relationships between cities and citizens in the Arctic. Thereby, the book's empirical and conceptual contribution adds to the emerging debate on human activities in the Arctic region that contribute to climate change and its complex effects in and beyond the region.

While urban development is a pressing issue across the circumpolar Arctic, this book focuses specifically on cities located in the European Arctic. The countries in Northern Europe (including Iceland and Greenland) have comparable governance structures and are closely linked economically to each other and with the European Union (EU), either as member states (Finland, Sweden), as members of a free trade association (Iceland, Norway through the European Economic Area), or as an Overseas Country and Territory or by special bilateral treaties (EU-Greenland fisheries agreement). The diversity of these countries' citizens (including Indigenous peoples),[19] economic and social challenges and opportunities, their legal frameworks and the infrastructure they provide in their Arctic urban centres allows a comparative overview of their different governance approaches to sustainable urban development and towards participation (cf. Chapter 3). These countries' local governments represent one, so-called "Nordic model" of local self-government (Hendriks *et al.* 2010, see Chapter 5). With this special empirical focus on the European Arctic, our book contributes to a better understanding of the potentials and limitations not only for implementing the 2030 Agenda, the Paris Agreement, and the New Urban Agenda as global visions but also for pursuing the regional visions agreed upon by the Arctic Council and the EU in the context of sustainable development.

In this book, we examine urban development approaches by focusing specifically on policy-making at the local level. More precisely, we concentrate on cities, which we define on the basis of an urban system-approach brought forward by Gerald Mills (2007) as "a variety of settlements of varying size and extent that are connected via transport and information corridors along which people, goods and information flow", which are embedded in a "national and international network of flows" (p. 1850). Similar to Mills, we further consider cities as agents, causes, and solutions of global change but acknowledge that cities are not homogeneous entities[20] and that they have different spaces to manoeuvre (see Chapters 4–7)—also in the European Arctic. Generalising findings is thus possible only to some extent (see Chapters 7 and 8). This book, however, provides insights on seven cities located in the European Arctic: Akureyri (Iceland), Kiruna and Luleå (Sweden), Nuuk (Greenland), Kolari and Rovaniemi (Finland), and Tromsø (Norway). We selected these cities based on three main drivers of change, which require political action at the local, national, regional, and global levels:

1 In the cities selected, the population has significantly changed (increased or decreased) in the past two decades (about 8% or more) or large economic activities and/or infrastructure measures are planned and about to be implemented soon, which will likely impact the city's overall population in the near future, the housing needs and access to basic services (see SDG 11, target 1).
2 The economic focus (e.g., tourism, shipping, mineral resources) of the cities differ as do their respective infrastructural needs, which require strong national and regional development planning (see SDG 11, target 11.8).
3 All cities share a certain affectedness and vulnerability to impacts induced by climate change as being located in the European Arctic, which require political action to reduce negative impacts of climate change on citizens and to reduce the environmental impact of the cities themselves (see Paris Agreement and SDG 11, targets 11.5, 11.6, 11.9).

While we did not investigate the respective approaches in details, these criteria helped us to identify cities that are affected by significant changes and allow us to compare the extent to which local approaches refer to related policies at other governance levels and elsewhere.

1.4 Selected Case Cities

In *Finland*, we selected Rovaniemi and Kolari as case cities. Rovaniemi is a main urban centre in Northern Finland and the capital of Finnish Lapland. With regard to the cities' economic and infrastructure development, Rovaniemi and Kolari have both been under consideration by the Finnish Transport Agency to be part of a Finnish-Norwegian cooperation to build and expand on a railway connection in the north-western Arctic. Considered for the rail link with the rest of Finland were also the ports of Murmansk (Russia) or Kirkenes (Norway) (using a connection via Rovaniemi) and the ports of Tromsø or Narvik (both Norway). Particularly with a view to Chinese plans to increase the use of Arctic shipping routes as a part of its "Belt and Road Initiative", the importance of these connections as a gateway to European markets is likely to increase. In addition, in Kolari, the change from a former large iron ore extraction industry to a popular skiing resort and tourism centre in the North promised an additional perspective on its urban development.

As we show throughout this book, Rovaniemi can be seen as "a city in tensions". While the "capital city of Lapland" is a strong, recognised brand, there are tensions between citizens and the city's administration on how the city should develop. Rovaniemi has a small city centre but many city districts, hosts the University of Lapland and offers wide-ranging public services. The city attracts a high number of guest workers and tourists each year who mostly come to visit Santa Clause Village. As a consequence of growing tourism, prices for housing are exploding. Since February 2022, the number of soldiers who are based close to Rovaniemi to protect the Finnish North has significantly increased, and soldiers are also more visible in the city itself. Rovaniemi is well connected (with many direct flights from and to other places in Europe and Asia), and the region of Lapland has a regional office in Brussels. Our interviewees described politicians from Lapland as very active at the EU level.

We perceived the Finnish city Kolari, on the other hand, as "a city shaped by societal conflicts". The city is sparsely populated: it spans more than 100 kilometres from the city's North to its South and a lot of infrastructure and services needs to be maintained (schools, transportation) for a comparatively small number of people. In geographic and societal terms, the city appears to be divided. Many buildings are isolated, and the city offers limited community meeting space. The Ylläs area in Kolari's North has developed as a tourist area, hosting one of Finland's most popular ski resorts and the Pallas-Yllästunturi National Park. Closer to Kolari's centre, in Hannukainen village, there is an iron–gold–copper mine. Mining was operated in the Hannukainen Mine from 1969 until 1989. The potential reopening of the mine has resurfaced various emotions for the city's citizens. On the one hand, the mine is considered crucial for local development and for service provision. The reconstruction of the railway would particularly improve the municipality's accessibility. On the other hand, the mine is expected to intensify societal challenges: It will provide employment opportunities, but most employees will be seasonal workers from distant places who will likely move to Kolari only temporarily for work purposes and without their families. Thus, they will not help the municipality to diversify and to reverse the demographic trend. Kolari has traditionally been Sámi territory but Sámi citizens moved to the outskirts of the town to maintain their traditional activities. Those engaged in traditional activities such as reindeer herding will, however, be greatly affected by the operation and infrastructure of the mine. Also, tourist activities in Kolari's North are perceived as being negatively affected by the re-opening of the mine. Due to a lack of trust and gridlocked positions, inclusive participatory discussions on the re-opening of the mine have not taken place.

In *Greenland*, we focused on the capital Nuuk, which is also its largest city. Greenland's overall population fluctuates: the natural increase is countered by ongoing outbound migration. Greenland's economy is dependent on fisheries and heavily supported by a fixed block grant from Denmark due to Greenland's status as an autonomous territory of the Kingdom of Denmark. Greenland's economic prospects with regard to the exploitation of resources (gold, rare earths, as well as uranium and offshore hydrocarbons) could be key for the country's economic diversification. At the same time, the influx of a work force from abroad could impact the local composition of the population (e.g., with regard to cultural background, age, gender) and intensifies urban development processes. Our interviewees often referred to the overruling power of the national government and its interest to support tourism and Greenland's economic independence, which is, however, seen as threatening Greenland's sensitive environment. An important element for the development of Nuuk is the fact that it is the capital of the country, which not only has positive effects; the city's authorities sometimes have limited influence on urban issues, which are ignored or subordinated to national interests and are within the decision-making circle of the national government. Issues of political (in)dependence, the colonial past, and ethnic tensions were also important themes in our research.

In *Iceland*, we investigated Akureyri, which is the second largest urban area in the country and home to a university, which is part of the University of the Arctic (UArctic) network. Overall, Iceland saw a rapid increase in tourism before the COVID-19 pandemic—the number of foreign passengers to Keflavik Airport (Reykjavík) almost quadrupled between 2010 and 2016. This massive increase also has had impacts on Akureyri with its large skiing resorts. The city's growth is part of a broader debate about the relationship between the capital region or so-called Greater (Stór)-Reykjavík and the rest of the country. This includes discussions about dispersal in investment or human capital flows and also the strategic topics related to the management of tourism in Iceland. Additionally, Akureyri presents an interesting case regarding the social and language integration of migrants into local communities and their participation, or lack thereof, in discussions on urban development.

In *Norway*, the research team focused on Tromsø, which is the largest city in Arctic Norway. It is home to several research institutions such as the Arctic University of Norway with one of its campuses in Tromsø. Further, the city hosts the secretariat of the Arctic Council, the Council's Indigenous Peoples' Secretariat, as well as the Arctic Economic Council's Secretariat and the Secretariat of the Arctic Mayors' Forum, which is also why Tromsø is also branded as "the capital of the Arctic". Based on our investigations, we see Tromsø as a city in transition. The city's potentially strong involvement in the expansion of Arctic shipping and resource transport is likely to lead to further urban development. Tromsø is the largest city in the European Arctic, and the municipality extends about 1,000 kilometres. It has a large Sámi population and citizens from 138 countries.

In *Sweden*, the cities of Kiruna and Luleå were chosen as case studies. Kiruna is a multi-cultural municipality with Sámi-, Finnish- and Swedish-speaking citizens located in Sweden's northernmost and largest municipality of the same name. The city is a prominent example of the interdependence between resource extraction, urbanisation, and colonisation. In its direct neighbourhood, the world's largest underground iron ore mine changes the city's landscape significantly. About a third of the city's inhabitants are currently being relocated to a destination about 3 kilometres east of the city's current centre. The move—organised and paid for by the Swedish state-owned mining enterprise LKAB—has already begun and is planned to be completed in 2035. While the mine is perceived as ensuring prosperity and providing services for Kiruna's residents, LKAB's expansion requires the relocation of Kiruna. By 2035, approximately 6,000 residents will have moved to new houses and town buildings. Moreover, at the beginning of Sweden's EU Presidency of the Council in January 2023, the discovery of Europe's largest-known so far rare earths deposit was announced in this area and framed as an important contribution to the implementation of the European Green Deal, which will further affect land use in the region.

Luleå is located in Sweden's northernmost county Norrbotten and is of particular interest with regard to the "iron ore line", a railway track that connects the harbour of Narvik in Norway via Kiruna with Luleå. Increasing resource development in Sweden's northern regions could directly impact the development of these cities. Luleå is home to a large ore harbour and hosts the Luleå University of Technology. The city also hosts a central data centre for Facebook in Europe. Thus, this case study gives insights into the impacts of energy intensive data management and data storage services on urban development and with regard to attracting new economic opportunities for European Arctic cities. We perceived Luleå as a vibrant city, developing in demographic and economic terms. This development is driven by convenient access to cheap renewable energy. This development impacts urban planning in the city, however, new investments (e.g., in factories) seem to be superseding citizens' social needs.

To systematically explore why local approaches to sustainable urban development differ in the European Arctic, we developed the conceptual model "Key drivers and factors shaping local approaches to sustainable urban development" (for a detailed explanation of the model see Chapter 2). With the help of this model, we explore and explain differences in the seven cities' local approaches to sustainable urban development. We identify imaginaries and cooperation as key drivers for how sustainable urban development is approached in the European Arctic, and we show how these two key drivers are embedded within global power structures (see particularly in Chapters 2, 5, and 7). Power imbalances, as we illustrate, are further enforced by limited connectivity, which reproduces centre-periphery relations and is one main barrier for aligning political priorities at the local level with those agreed upon at the national and regional levels.

Our results emphasise in particular the human dimension in policy-making on sustainable urban development. While remoteness is a factor complicating cooperation, collaboration, and policy-alignment, it is not the only one. Instead, the seven cities investigated here illustrate that it is remoteness in combination with other varying factors (lesser so in the cases of Nuuk and Luleå), which depend on the city context, that challenge the pursuit of the global goals across governance levels and regions in the European Arctic. The establishment of inclusive participatory processes and the transfer of knowledges across governance levels and regions particularly seem most demanding amongst others because of a lack of formalised structures: irrespective of the location, size, and infrastructure, we were surprised to discover that in all places studied, knowledge exchanges geared towards achieving the global goals were perceived to be limited. This applies not only to knowledge exchanges across governance levels (vertical dimension) but also to knowledge exchanges at the local level (horizontal dimension). Consequently, the places under investigation can be seen as being little engaged in how the global goals are (re-)defined at the national level and often "local entrepreneurs", individuals with well-established networks also beyond the local level, seem to shape the interpretation of changes and the focus areas in visions of the cities' future. Regulations on how participation (actor specific dialogues, open hearings and consultations) should be organised and at what point are often vague or do not exist even if participation is mandatory in urban planning processes. In that regard, the smaller cities investigated in particular have difficulties organising meaningful dialogues due to a lack of capacity, which is why participatory tools often seem to fill legal requirements instead of encouraging in-depth exchanges on content. Against this backdrop, it comes as no surprise that urban cooperation appears to be much more limited than we expected and that the (missing) success of collaboration is rooted in individual relationships and not in coordinated national or regional strategies.

1.5 Structure of the Book

To explore the question of why visions for sustainable urban development are not aligned in remote regions, we investigate sustainable urban development approaches in this book in three steps: first, we explore how national and local legislation and governance attempt to implement the global visions (Chapter 3) and how their application shapes participation in urban development processes in the European Arctic (Chapter 4).[21] Second, we investigate how sustainable urban development approaches can be better aligned through transnational cooperation in theory (Chapters 2 and 3) and practice (Chapter 5).[22] Third, we shed light on how the engagement of different state and non-actors can be enhanced to advance policy alignment in the context of sustainable urban development (Chapters 6 and 7).[23]

Chapter 2 sets off by discussing how sustainable urban development is envisioned in global instruments. First, we illustrate the extent to which multi-level and multi-actor approaches shall carry forward the pursuit of the global goals. The governance instruments describe multi-level governance and multi-actor approaches as enabling governments and actors across policy-fields and governance levels to pursue the goals agreed upon, amongst others, by taking advantage of synergies and providing access to information and means of implementation, including finances. The global agreements do not relate to the challenges that arise from pursuing both approaches and neglect how power differentials shape coordination and cooperation across governance levels and among state and non-state actors. Second, we show how multi-level and multi-actor approaches are reviewed in governance studies and in sustainability and urban studies. The literature on polycentric and indirect governance underlines the different type of linkages among actors that shape their cooperation and the dynamic political orders in which multi-actor partnerships operate. As a consequence, leadership takes different forms as collaboration is steered by different actors at different levels (see the concept of orchestration, Chapter 2). Third, we illustrate how studies from the fields of sustainability and urban studies assess the related challenges in collaboration that result from differences in organisational settings, institutional logics, high levels of uncertainty, and governance complexity. Based on these insights and fourth, we then recognise the importance of multi-level governance and orchestration for pursuing the global goals and develop the conceptual model "Key drivers and factors shaping local approaches to sustainable urban development" that considers different issues of relevance for sustainable urban development in the Arctic. With this model, we argue that it is possible to develop pathways for how the implementation of global visions such as the Paris Climate Agreement, the 2030 Agenda, and the New Urban Agenda can be advanced in remote regions, notably, the European Arctic is an example.

In *Chapter 3*, we show the diverse landscape of governance on the international, national, and municipal/city level reflect how the three components are embedded in legal and governance frameworks, which we consider determining factors for how sustainable urban development is approached at the local level (actors and their relationships, institutions and their set-ups, political priorities). This chapter thus complements the insights provided in the previous chapter by providing an overview on national legislation and policies and the local specifications of implementation in our case cities. First, we explore the legal implications of the three global agreements (the 2030 Agenda, the Paris Agreement, and the New Urban Agenda) and provide an overview on their legal implications and key contents on participation that ought to be considered in national legislation and policies for the implementation of the global goals. Second, we shed light on the legal framework at the national level in which our case cities can potentially shape policy approaches to sustainable urban development. Third, we relate to our case cities and provide an overview of their local engagement in the implementation of the global goals in the context of sustainable urban development.

Chapter 4 then "zooms in" on the Nordic model of governance and identifies the prerequisites and obstacles to participatory approaches. The chapter first introduces the Nordic model of governance and second, reviews it under consideration of the governance practices in our case cities. Thereby it identifies a gap between governance practices and the assumptions of the Nordic Model, more specifically between the formal and the informal rules of governance. The chapter further shows how participatory processes are also weakened by the ambiguity of the concept of sustainable development, which seems to be mostly structured to maintain the status quo instead of developing innovative pathways.

Building on the provided insights from the local context in Chapters 3 and 4, the subsequent *Chapter 5* "zooms out" to investigate how local approaches in the context of urban development are embedded in cooperative frameworks beyond national contexts. Given the scholarly debate on transnational cooperation in the Arctic and the insights shared by our interviewees, we show that urban cooperation beyond national contexts has been less intense and extensive than we expected. This finding suggests that at the transnational level, cooperation between our case cities has not contributed to policy-alignment in the context of sustainable urban development and is not (yet) orchestrated by national and regional authorities. Moreover, cooperation between cities in the European Arctic does not seem to be driven by the global goals but by selective local priorities. This chapter also discusses the significant impact of external crises in recent years (the COVID-19 pandemic and the international crisis caused by Russia's full-scale invasion of Ukraine on transnational cooperation in the European Arctic).

Based on the findings presented in the preceding chapters, in *Chapter 6* we introduce pathways and scopes for adjustments for pursuing sustainable urban development in the European Arctic. First, through the lens of our conceptual model, in this chapter we reflect upon the indications provided by our empirical findings on why urban development approaches in the European Arctic are not aligned with the global goals. Second, we sketch three pathways for how local approaches to sustainable urban development can be better aligned via (1) a reformation of the Nordic Model, (2) more inclusive and active multi-actor platforms, and (3) a regional approach.

While the preceding chapters focus specifically on the European Arctic, *Chapter 7* broadens the geographical perspective to remote regions in general. First, it compares the challenges and opportunities for pursuing sustainable urban development in the European Arctic with those identified in other remote regions. We show that in other remote regions as well policy alignment across governance levels and participatory approaches are considered weak. Second, based on a detailed comparison with our cases, we discuss the implications for the scholarly debate on sustainable urban development in remote regions more generally. Third, we critically discuss the transferability of our model, identify differences and highlight gaps of knowledge. We conclude that despite all limitations that come along with the place-sensitivities of cities and that we also observed in our empirical data, our model grasps various similarities between the European Arctic and other remote regions and thus seems to be applicable also beyond the European Arctic.

The final *Chapter 8* summarises and assesses the main findings of the research presented in this book. It outlines the challenges and opportunities for pursuing the global goals and shared visions in the context of sustainable urban development in the European Arctic and highlights the significant gap between the aspirations of the global agreements and the realities, we investigated in our seven case cities. In this chapter, we also present future research directions and nine policy recommendations for fostering sustainable urban development in the European Arctic.

1.6 Methodology and Positionalities

The purpose of this book is to provide a better understanding of the opportunities and challenges for advancing cooperation and policy-alignment in the European Arctic and in other remote regions to pursue the global goals. Our research was driven by two exploratory questions:

1 How can sustainable urban development be advanced in remote regions?
2 How can sustainable urban development policies be aligned across levels and based on the hypotheses that urban development will be more sustainable

  a if the perspectives of local actors are reflected in policy making and decision making and
  b if policies correspond to another across governance levels.

Our research was oriented on real-world practices. The methodological ground of the research presented in this book is shaped by the pragmatic paradigm and by the research paradigm of social constructivism. In accordance with the pragmatic paradigm, our research insists on the practical effectiveness of solving research problems and postulates a pluralist approach to the issues examined (Creswell and Creswell 2022). By following the research paradigm of social constructivism, which highlights the social construction of politics, we consider the multiple meanings that are ascribed to the object under investigation (the implementation of the SDGs at the city level), integrate different research angles (interdisciplinarity) and perspectives from researchers and practitioners and triangulated our data respectively (Hay 2015, Wendt 1992). To explore questions at the interface of disciplines, the researchers in the team worked closely together in all research phases. These interactions allowed for the consideration, discussion, and alignment of diverging perceptions.

In consideration of the research paradigms, our assessments are also shaped by the researchers' interdisciplinary interpretations and positionalities. As Polish and German researchers trained in the social and legal sciences, we are taking up a double "outsider" position (Toy-Cronin 2018) in Arctic research both in terms of citizenship/place of residence and disciplinary background. As non-Arctic citizens, we were also geographically distant to the region under investigation. However, in times of transboundary challenges and the relevance of the Arctic for the world's climate, we share an intrinsic research interest in the region and its social-political fabric.

The results presented in this book are based on a sequential exploratory qualitative research design. More specifically, in addition to a critical reading of the available scientific literature, the research team conducted (1) a qualitative content analysis[24] of relevant policy and legislative documents (i.e., local planning documents and laws, directives) and official statistics, as well as (2) semi-structured interviews with key informants from different actor groups (residents; representatives of the administration, of civil society organisations, of businesses; elected officials and researchers).[25] The team further considered (3) place observations from research visits in the remote cities under investigation.

The data collection phase started in December 2021 and continued until August 2023. All researchers conducted fieldwork in teams, following the purpose to take advantage of different disciplinary perspectives and of complementary skills (such as different conceptual angles, empirical knowledges, and accesses to scientific networks). The researchers involved created city-teams based on three main criteria:

1 To provide equal access to data and to assess findings jointly, each city-team was composed of researchers from Poland and Germany.
2 The city-teams included researchers from different work packages.
3 To facilitate in-country comparisons in countries with two case cities, the same researchers focused on both cities.[26]

To limit research biases from the researchers, interviews were prepared and carried out mostly and whenever possible in pairs. The team used purposive sampling and snowballing to reach key informants. The interviews were conducted in English, virtually and in the cities under investigation between May 2021 and August 2023, in a semi-structured, loose conversational manner by touching on standardised questions shared with the interviewees ahead of the meeting. Each interview lasted 45 to 60 minutes and consisted of questions about the meaning and practices of sustainable development in and beyond the cities under analysis (cf. questionnaire provided in the Annex). Also, the COVID-19 pandemic and the war in Ukraine were added to the set of questions given their impact on the topic under investigation. Ahead of the interview, all interviewees received and agreed to a consent form, and the majority of interviews were recorded and transcribed. Both researchers took notes in the cases where interviewees did not agree to recording the interview. To enable interviewees to speak more openly about sensitive issues, the researchers promised all interviewees anonymity. Insights from 80 interviews were supplemented by place observations in all cities under analysis except for Nuuk (particularly due to travel limitations during the COVID-19 pandemic). All transcripts and notes were anonymised and reviewed by the city-teams.

Due to the COVID-19 pandemic, the related travel restrictions and the political dynamics caused by the Russian invasion in Ukraine, the team had to adapt the data collection phase. The team started this phase with desk research and virtual interviews. In the first months after the Russian invasion of Ukraine, however, it was more difficult to find interviewees willing to participate in virtual interviews due to security concerns and the frequent experience with spyware. Since questions on sustainable urban development intersect with energy security, our research focus became a more politicised one with the Russian war against Ukraine. To still reach at least two representatives from each actor group in the cities under analysis and to capture potential diverging perceptions, the team extended this phase longer than initially planned. "Going virtual" instead of visiting the cities under investigation required the team to develop a new toolkit for establishing contacts (amongst others, the team shared a project flyer and a video on the research and on the researchers in the letters of invitation to build trust). To better understand the circumstances in the cities under investigation without being able to visit them, the project team organised a hybrid expert workshop with various researchers from the case countries under analysis. The workshop served as a sounding board for the project, in which 40 researchers participated and discussed preliminary findings and the research design. Due to the travel uncertainties, the team followed a flexible approach and travelled to the cities under investigation whenever travel restrictions allowed so. It was not possible to organise the intended focus group meetings in the cities under analysis, which would have required early and reliable planning.

The phase of data analysis started in October 2021 and terminated in August 2023. Based on the project's goals and the questionnaire, the team then developed an initial code system and defined coding rules. After all researchers had tested the initial, inductively derived code system, the team adapted and extended the code system and the coding rules deductively to also include codes on unexpected insights and new topics from the data. In a next step, the city-teams coded all data for the respective cities under analysis by using the software MAXQDA. The teams applied structural coding to compare answers from interviews and simultaneous coding (coding of data with multiple codes) to grasp different layers of data. To minimise the risk of confirmation bias (such as frequency illusion) and selective attention, the coding was carried out in pairs. We followed a sequential coding process (one person after another) because a high score of inter-coder reliability (achieved if researchers discuss and agree on how to code the same data) was not of relevance for our research design or the research questions. As the city-teams were composed of researchers working on different work packages, they considered different questions during the coding process. The data was coded with multiple codes depending on the perspectives of the project's work packages. Based on the thematic analysis of the transcripts and the notes taken by the researchers, the city teams prepared case studies for each city under investigation and identified recurring patterns, themes, and categories. The research team then discussed their analyses of the different cases together during a workshop to identify differences and similarities, unexpected observations, contradictory data and inconsistencies (negative case analysis), limitations, and demands for further research.

Based on the subsequent cross-case analysis, the research team compared and evaluated the city cases to explain:

1 Why do approaches to sustainable development differ?
2 How can sustainable urban development in the European Arctic be steered more effectively in alignment with local and global policies?
3 To what degree is the European Arctic specific or typical for remote regions more generally?

The results of our analysis and evaluation are summarised in the following chapters of this book, in which we argue that the local level needs to receive more attention in multi-level governance and that local views will have a greater impact at the global level if they are aligned transnationally. Both need to be better understood and address local challenges when envisioning the achievement of the goals agreed upon in the 2030 Agenda for Sustainable Development and the Paris Climate Agreement.

1.6.1 Research Limitations

Despite the efforts and intention to reach interviewees and to include in particular the perspectives of those often less represented in the past and in ongoing research on city development in the European Arctic, the team faced challenges in conducting an equal number of interviews with representatives from all actor groups. The perspectives from representatives of civil society organisations and citizens not wearing "double-hats" (belonging also to one of the other group of actors under analysis) in particular are under-represented in our samples. Moreover, the total number of interviews conducted in each city differs (Table 1.1).

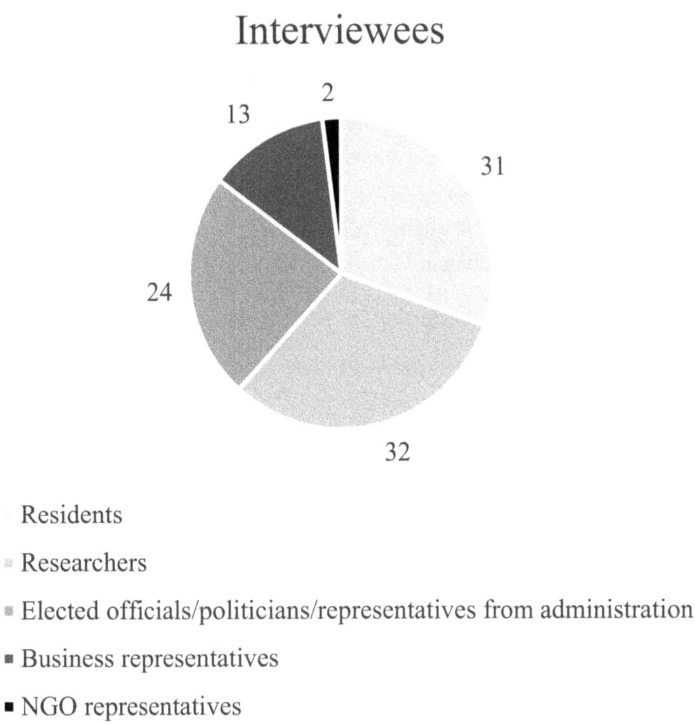

Figure 1.1 Perspectives considered in interviews. Own work.

Table 1.1 List of interviews

City	*Interviewee*	*Type of interview*	*Date of interview*
Akureyri	Researcher	Online	06.07.2022
Akureyri	Researcher	Online	27.05.2022
Akureyri	Researcher/resident	Online	06.07.2022
Akureyri	Researcher/resident	Online	15.09.2022
Akureyri	Representative of the administration	In person	14.10.2022

(*Continued*)

Table 1.1 (Continued)

City	*Interviewee*	*Type of interview*	*Date of interview*
Akureyri	Representative of the administration	Online	09.05.2023
Akureyri	Elected official	Online	24.10.2023
Akureyri	Business representative	Online	12.04.2023
Akureyri	Business representative	Online	26.04.2023
Akureyri	Business representative	Online	28.04.2023
Akureyri	Resident	Online	13.07.2022
Akureyri	Resident	In person	11.10.2022
Akureyri	Resident	Online	26.04.2023
Luleå	Researcher/resident	Online	16.12.2021
Luleå	Researcher/resident	Online	20.01.2022
Luleå	Researcher/resident	Online	20.12.2021
Luleå	Researcher/resident	Online	05.04.2022
Luleå	Representative of the administration	Online	26.10.2022
Luleå	Elected official	Online	03.05.2022
Luleå	Elected official	Online	30.09.2022
Luleå	Elected official	online	08.05.2023
Luleå	Business representative	Online	31.03.2022
Luleå	Business representative	Online	01.04.2022
Kiruna	Business representative	Online	16.03.2022
Kiruna	Representative of the administration	Online	18.01.2022
Kiruna	Researcher	Online	21.03.2022
Kiruna	Researcher	Online	20.01.2022
Kiruna	Researcher	Online	16.12.2021
Kiruna	Researcher	Online	20.12.2021
Kiruna	Researcher/representative of civil society organisations	Online	17.01.2022
Kolari	Resident	In person	21.11.2022
Kolari	Resident	In person	21.11.2022
Kolari	Resident	In person	21.11.2022
Kolari	Resident	In person	21.11.2022
Kolari	Resident	In person	21.11.2022
Kolari	Researcher	Online	28.01.2022
Kolari	Researcher/resident	Online	13.01.2022
Kolari	Politician	In person	21.11.2022
Kolari	Politician/business representative	Online	07.03.2022
Kolari	Elected official/business representative	Online	07.03.2022
Kolari	Researcher/resident	Online	26.01.2022
Kolari	Former elected official/business representative	Online	04.03.2022
Nuuk	Researcher	Online	29.04.2022
Nuuk	Researcher	Online	04.08.2022
Nuuk	Researcher	Online	15.07.2022

(*Continued*)

Table 1.1 (Continued)

City	*Interviewee*	*Type of interview*	*Date of interview*
Nuuk	Representative of the administration	Online	31.03.2022
Nuuk	Representative of the administration	Online	04.04.2022
Nuuk	Representative of civil society organisations /resident	Online	29.04.2022
Nuuk	Researcher/representative of the administration	Online	16.08.2023
Rovaniemi	Resident	In person	20.11.2022
Rovaniemi	Resident	In person	20.11.2022
Rovaniemi	Resident	In person	20.11.2022
Rovaniemi	Resident	In person	20.11.2022
Rovaniemi	Resident	In person	20.11.2022
Rovaniemi	Resident	In person	22.11.2022
Rovaniemi	Business representative	In person	20.11.2022
Rovaniemi	Researcher	In person	22.11.2022
Rovaniemi	Researcher	Online	28.01.2022
Rovaniemi	Business representative	Online	18.02.2022
Rovaniemi	Former elected official	Online	07.02.2022
Rovaniemi	Elected official	Online	02.02.2022
Rovaniemi	Elected official	Online	24.11.2022
Rovaniemi	Researcher	Online	27.05.2021
Rovaniemi	Researcher	Online	20.01.2023
Rovaniemi	Business representative	Online	09.03.2022
Rovaniemi	Researcher and politician	Online	13.05.2022
Tromsø	Business representative/resident	Online	25.04.2023
Tromsø	Researcher/resident	Online	15.03.2023
Tromsø	Politician/resident	Online	20.03.2023
Tromsø	Elected official	Online	20.02.2023
Tromsø	Researcher/resident	Online	08.03.2023
Tromsø	Researcher/resident	Online	08.03.2023
Tromsø	Researcher/resident	Online	13.03.2023
Tromsø	Administration/resident	Online	24.03.2023
Tromsø	Elected official	Online	10.03.2023
Tromsø	Politician/researcher/resident	Online	27.03.2023
Tromsø	Researcher	Online	17.03.2023
Tromsø	Researcher	Online	08.03.2023
Tromsø	Researcher	online	26.01.2023

Reasons for these differences are:

1 Language-barriers: All interviews were conducted in English, which is not the first language in the cities under investigation.
2 The ratio of people: In small cities (such as Kolari), the number of people engaged in the topic under investigation and with capacities to do research interviews is smaller than in larger cities (such as Tromsø).

3 Interviews conducted before and after the Russian invasion in Ukraine: Our research addressed questions on infrastructural development and political cooperation during a politically sensitive time. Some informants declined to speak with us due to the sensitivity of the topic. Furthermore, the priorities of public administrations shifted due to the impacts of the war on local developments.

While doing virtual interviews provided more equality in regard to the research setting, as we met with informants "in the same place", we noted that interviews were also more formal, and it was more difficult to establish a trustful atmosphere and personal connection with informants that we had not met before. Further research—also the envisioned focus group interviews—will be needed to substantiate the data that we considered in our analyses accordingly (Figure 1.1).

Acknowledgements

This introductory chapter was written by Dorothea Wehrmann. Jacqueline Götze drafted the section Methodology and Positionalities, and reviewed the chapter together with Michał Łuszczuk and Arne Riedel. The chapter was further reviewed by three external experts. All authors of this book read and accept the content of this introductory chapter.

Notes

1 Cf. Social cognitive learning theory (i.a. Rumjaun and Narod 2020, Whitham *et al.* 2021).

2 Collaboration can be defined as "the process of working together to develop and sustain the solution of shared problems" in contrast to cooperation, which "occurs when participants agree on a shared problem that they try to solve through a division of labour" (Chaturvedi *et al.* 2021, p. 16).

3 There is no clear-cut definition of imaginaries because of "their largely unstructured, unlimited, and indefinite nature" (Buitrago *et al.* 2016, p. 6). However, the understanding of imaginaries as lenses is influenced by research in disciplines ranging from History and Philosophy to Sociology and builds on the study of imaginaries, amongst others, by Cornelius Castoriadis, Émile Durkheim and Charles Taylor. For more details, see Chapter 2.

4 The sustainability-paradigm was pushed forward by the Brundtland Report issued by the World Commission on Environment and Development in 1987 (UN 1987). Following the report, sustainable development "meets with the needs of the present without compromising the ability of future generations to meet their own needs". It builds upon a concept rooted in sustainable forest management (only cut as many trees as can regrow in a certain period of time). With the adoption of the Agenda 21 at the UN Conference on Environment and Development (UNCED) in Rio de Janeiro 1992, the concept of sustainable development became institutionalised (Petrov *et al.* 2017, p. 7). Then, at the 2002 World Summit on Sustainable Development "three 'interdependent and mutually reinforcing pillars' or dimensions of sustainable development: economic, social, and environmental" (Pram Gad *et al.* 2019, p. 4) were introduced. The social dimension, for instance, entails

human development and equity, which "implies a fair distribution of benefits and a discourse that has its roots in tensions related to colonialism and decolonization" (Petrov *et al.* 2017, p. 3). Overall, "sustainability emerged as a global concern in way that was politically programmatic before it was academic" (Pram Gad *et al.* 2019, p. 1).

5 Progress on the global targets agreed upon in the agreements is reviewed on a country-by-country basis by the High-level Political Forum of the United Nations Economic and Social Council, the United Nations Framework Convention on Climate Change (UNFCCC) and the United Nations Human Settlements Programme. While ideally the concept allows for comparisons of development status between units at different points in time and facilitates learning and cooperation, development approaches and measurement and monitoring frameworks are controversial (Mawdsley 2021, de Mello e Souza 2021).

6 Green Colonialism relates to imperialistic efforts deployed to control the exploitation of nature by actors outside their own national jurisdictions.

7 Accordingly, in our interviews, we did not provide a definition of this concept. Instead, we invited our interviewees to share their understanding of the concept with us to explore how much these differed.

8 In 2014 Arctic Human Development Report identified "Arctic settlements, cities, and communities" as one of the main gaps in knowledge of the region (Larsen and Fondahl 2014, p. 24–25).

9 Similar to the framing used in the global visions, Svennevig convincingly argues that "governments and other stakeholders have learnt that it is necessary to activate the local population in and around the affected areas if nature conservation is to succeed" but also emphasises that the category "'Locals' introduces an actor who is not necessarily interested in nature conservation", which affects the collaboration and relates to the debate on "who has the right to participate as 'locals'—and when in the decision-making process they should be involved" (Svennevig 1997, p. 2; machine translated from Danish to English by DeepL). As we show in the subsequent chapters, the dichotomy between "Indigenous" and "non-Indigenous" citizens further complicates the definition and inclusion of "local perspectives" in policy- and decision-making. While our research relates to "local approaches" as approaches that are shaped by the people who live in a particular place (e.g., in the cities under analysis) and thus by geographically determined borders, we do not neglect that (1) perspectives among "locals" most likely differ and that "locals" can also be people who feel a special sense of belonging without living in the places under analysis and (2) nature cannot be managed as determined by geographically determined borders only but rather needs to be addressed in light of thematic issues (e.g., fishing activities affect also other places beyond geographic conditions) and accordingly needs to be considered in broader perspective (Svennevig 1997, p. 3 and 10).

10 For Indigenous peoples, the Arctic has always been a transnational space. With the end of the Cold War also the Arctic-rim states considered the Arctic as "a common space" (Knecht and Keil 2013, p. 22). Since then, many environmental, societal, and economic concerns in the Arctic have been perceived as being of "transnational nature" that needed to be addressed via regional approaches. This regional cooperation, particularly under the auspices of the Arctic Council, where Indigenous peoples hold special rights and are represented as permanent participants, has often been reviewed as unique in global governance (Wehrmann 2020).

11 Defined as interactions among actors from different actor groups (including at least one non-state actor) that occur on a regular basis, cross borders but are not global in scope (Albert *et al.* 2009, Pries 2010).

12 Remoteness can be explored in multiple ways. Until present, there is no universal definition nor does a remoteness index exist (Stringer *et al.* 2023).

13 In general, remote regions are far distant from markets and centres of business (geographic remoteness), which causes a separation of producers and consumers (economic remoteness). This leads to increasing transportation costs and overall higher business costs in remote regions. In turn, less economic activities take place and costs of living are higher due to scarce infrastructure. The people's mobility is limited due to higher transportation costs, time constraints and unsafe travel options.

14 The Arctic has often been portrayed as an exceptional governance space (Spence *et al.* 2023) because of the long experience of peaceful cross-border cooperation that has shaped the region despite of its geopolitical relevance.

15 This relates particularly to the provision of public services, which is a central challenge in remote regions due to the limited infrastructure in large territories. With expenses for public goods being usually high, access to higher education not given in every city, and high expenses for mobility, remote regions—such as the European Arctic—experience outward-migration and brain drain spurring demographic change.

16 A prominent example is the contested wind power plants in Northern Norway (the Fosen Peninsula), due to the historic decision of the Norwegian Supreme Court (two years ago), which ruled against the Norwegian Ministry of Petroleum and Energy because the windfarms are located in an area where reindeer husbandry is practiced and, thus, interfered in the traditional Sami areas (Norwegian Human Rights Institution 2023). This example illustrates that the people from the North increasingly use legal means to go against decisions from national governments.

17 For a detailed analysis of the "ambiguous nature of responsibility as a normative element of global governance", see Hansen-Magnusson (2019).

18 Canada, the Kingdom of Denmark (Greenland), Finland, Iceland, Norway, Russia, Sweden, and the United States.

19 There are different Indigenous peoples with traditional land in the European Arctic. The majority of the people of Greenland are Inuit, whereas parts of Norway, Sweden, and Finland belong to Sápmi, the homeland of the Sámi people. Although they are all Indigenous peoples, they form part of diverse communities, with some following traditional livelihoods such as fishing, reindeer herding, handicrafts, and hunting—to name just a few. Being Indigenous is not defined by traditional livelihoods though, but through the acceptance of the Indigenous communities themselves. In the case of the Sámi people, for instance, two-thirds live outside Sámi homeland (Joona 2020; Nyseth and Pedersen 2014).

20 In the circumpolar Arctic region, settlement patterns reflect the historical demographic and economic modes of development: dispersed settlements are based on local and decentralised harvesting of natural resources: "company" towns are centred around large companies being the major local employer (e.g., Kolari and Kiruna), whereas more diversified towns and cities include centres of local authority (municipalities), public services, and trade and transport hubs (e.g., Tromsø, Rovaniemi). The Arctic regions are thus settled in a rather contrasting way, with vast sparsely inhabited or uninhabited regions interspersed with few relatively big cities. As Laruelle points out, in the polar context, a city is defined by the fact that the people living in this settlement do not earn their living through agriculture or hunting, but through industrial, service, or administrative work and not necessarily by the presence of a certain number of inhabitants or a certain population density (2019). According to this definition, which excludes villages where traditional ways of life still predominate, there are about 60 cities in the Arctic region. The hinterlands in most functional urban areas in the Nordic region do include towns and suburbs as well as rural areas (Nordregio 2018). As Laruelle (2019) convincingly explains, the latest urbanisation wave in the Arctic has been driven by industrial activities, the militarisation of the Arctic and the development of regional administrative centres.

21 This was the guiding question of the project's work package one, investigated by Jacqueline Götze, Katarzyna Radzik-Maruszak and Arne Riedel. Subordinate questions

were: How do local and national legal and governance frameworks affect the participation of actors in policy-making processes that concern urban development in the European Arctic? What challenges go along with the implementation of these frameworks and are they reflected by indicators currently used? How could national and local legal frameworks be improved to enable and support participatory approaches effectively, and potentially be harmonised through transnational agreements?

22 This question guided research in the project's work package two, investigated by Michał Łuszczuk and Dorothea Wehrmann. Subordinate questions were: How can multi-stakeholder (cooperation formats among governmental, non-governmental, and private actors that are based on agreements) and participatory approaches (such as public consultations or referenda) that promote inclusive decision-processes be applied in the context of Arctic urban development? What challenges go along with their implementation? How and under what conditions can transnational cooperation stimulate and enhance bringing together different knowledges and perspectives in the context of sustainable urban development in remote areas? How can such forms of cooperation be organised and operationalised, e.g., by institutions such as the Arctic Council?

23 This question guided the research organised in the project's work package three, examined collaboratively by all authors of this book. Subordinate questions were: In how far are pathways developed for (specific areas in) the European Arctic useful for sustainable urban development in other regions? How can complex qualitative data on diverse and rapid changes and their multi-dimensional effects of urbanisation processes across remote regions such as the Arctic be quantified? In how far can the framework and data provided by the research project contribute to the implementation of related Sustainable Development Goals (SDGs)?

24 A systematic qualitative content analysis is a social sciences' method for analysing different types of text data, involving different coding processes to identify themes and patterns in text data (Hsieh and Shannon 2005, Mayring 2010).

25 Categorising our interviewees in these six actor groups helped us to better understand the diversity among local perspectives. These categories shall not imply that one actor group stands for one specific perspective or interest only. Within these actor groups, quite different perspectives were shared and interests put forward.

26 Kiruna and Luleå were investigated by Jacqueline Götze; Kiruna, Luleå, Nuuk, and Akureyri by Michał Łuszczuk; Kolari, Rovaniemi, and Tromsø by Katarzyna Radzik-Maruszak; Nuuk and Akureyri by Arne Riedel; and Kolari, Rovaniemi, and Tromsø by Dorothea Wehrmann.

References

Albert, M., *et al.*, 2009. Introduction: The communicative construction of transnational political spaces. *In:* Mathias Albert, G. B., *et al.*, eds., *Transnational political spaces. Agents—structures—encounters*. Frankfurt: Campus, 7–31.

Aleksandrov, E., and Dybtsyna, E., 2024. Smart cities for a sustainable Arctic? Introducing critical debate. *Polar Geography*, 47 (2), 1–21. https://doi.org/10.1080/1088937X.2024.2351496.

Balestreri, C., Kurucz, E. C., McIlwraith, T., and Jacobs, S., 2023. Local understandings and global challenges: Exploring sense of place in sustainability transitions. *Ecology and Society*, 28 (1), 33. https://doi.org/10.5751/ES-13895–280133.

Béland, D., and Cox, R. H., 2015. Ideas as coalition magnets: Coalition building, policy entrepreneurs, and power relations. *Journal of European Public Policy*, 23 (3), 428–445.

Berghöfer, U., et al., 2022. "Societal relationships with Nature": A framework for understanding nature-related conflicts and multiple values. *People and Nature*, 4 (2), 534–548. https://doi.org/10.1002/pan3.10305.

Berman, M., and Orttung, R. W., 2020. Measuring progress toward urban sustainability: Do global measures work for Arctic cities? *Sustainability*, 12 (9), 3708. Available from: https://www.mdpi.com/2071-1050/12/9/3708 [Accessed 13 February 2022].

Bertelsen, R. G., 2019. The Arctic as a laboratory of global governance: The case of knowledge-based cooperation and science diplomacy. *In:* Finger, M., and Heininen, L., eds., *The GlobalArctic handbook*. Cham: Springer International Publishing, 251–267.

Böhme, K., Zillmer, S., Hans, S., Hrelja, D., Valenza, A. & Mori, A., 2022. *Research for REGI Committee – The impacts of the COVID-19 pandemic on EU cohesion and EU cohesion policy - Part I: Overview and first analysis*, European Parliament, Policy Department for Structural and Cohesion Policies, Brussels.

Brand, A., Furness, M., and Keijzer, N., 2021. Promoting policy coherence within the 2030 agenda framework: Externalities, trade-offs and politics. *Politics and Governance*, 9 (1), 108–118. Available from: https://doi.org/10.17645/pag.v9i1.3608 [Accessed 21 December 2022].

Brenner, N., 2019. *New urban spaces: Urban theory and the scale question*. New York: Oxford University Press.

Brescia, R., and Marshall, J. T., eds., 2016. *How cities will save the world*. London: Routledge.

Brown, T., 2016. Sustainability as empty signifier: Its rise, fall, and radical potential. *Antipode*, 48 (1), 115–133.https://doi.org/10.1111/anti.12164.

Buitrago, C., Petersen, M., Tyrell, P.-M., and Wehrmann, D., 2016. Introduction: (Geo-) political imaginaries in the Americas. *Fiar. Forum for Inter-American Research*, 9 (1), 4–18. Available from: https://interamericaonline.org/wp-content/uploads/2016/05/01_fiar-Vol-9.1-Inroduction-Geopolitical-Imaginaries-4-18_v2.pdf[Accessed 30 April 2024].

Chaturvedi, S., *et al.*, 2021. Development cooperation in the context of contested global governance. *In*: Chaturvedi, S., *et al.*, eds., *The Palgrave handbook of development cooperation for achieving the 2030 agenda: Contested collaboration*. Cham: Springer International Publishing, 1–21.

Creswell, J. W., and Creswell, J. D., 2022. *Research design: Qualitative, quantitative, and mixed methods approaches*. London: SAGE.

De Mello e Souza, A., 2021. Building a Global Development Cooperation Regime: Failed but Necessary Efforts. *In*: Chaturvedi, S., *et al.*, eds., *The Palgrave Handbook of Development Cooperation for Achieving the 2030 Agenda*. Palgrave Macmillan, Cham. Available from: https://doi.org/10.1007/978-3-030-57938-8_16 [Accessed 24 February 2024].

Fuhr, H., Hickmann, T., and Kern, K., 2018. The role of cities in multi-level climate governance: local climate policies and the 1.5°C target, *Current Opinion in Environmental Sustainability*, 30, 1–6. Available from: https://doi.org/10.1016/j.cosust.2017.10.006 [Accessed 25 November 2024].

Gamble, J., and Shadian, J., 2017. One Arctic…But unevencapacity: The Arctic Council permanent participants. In: Lackenbauer, P. W., Nicol, H., and Greaves, W., eds., *One Arctic: The Arctic Council and circumpolar governance*. Ottawa: Canadian Arctic Resource Committee/Centre on Foreign Policy, 142–156.

Hannigan, J., and Richards, G., eds., 2017. *The SAGE handbook of new urban studies*. London: SAGE.

Hansen-Magnusson, H., 2019. The web of responsibility in and for the Arctic. *Cambridge Review of International Affairs*, 32 (2), 132–158. Available from: https://doi.org/10.1080/09557571.2019.1573805.

Haug, S. & Taggart, J., 2024. Global Development Governance 2.0: Fractured accountabilities in a divided governance complex. *Global Policy*, 15, 128–134. Available from: https://doi.org/10.1111/1758-5899.13261

Hay, C., 2015. Social constructivism. *In*: Bevir, M., and Rhodes, R. A. W., eds., *Routledge handbook of interpretive political science*. London: Routledge, 99–112.

Hemmersam, P., 2021. *Making the Arctic city. The history and future of urbanism in the circumpolar North*. London, New York, Dublin: Bloomsbury Publishing.

Hendriks, F., Lidström, A., and Loughlin, J., 2010. *The Oxford handbook of local and regional democracy in Europe*. Oxford, New York: Oxford University Press.

Hsieh, H. F., and Shannon, S. E., 2005. Three approaches to qualitative content analysis. *Qualitative Health Research*, 15 (9), 1277–1288. https://doi.org/10.1177/1049732305276687.

Joona, T., 2020. ILO Convention No. 169 and the governance of indigenous identity in Finland: Recent developments. *The International Journal of Human Rights*, 24 (2–3), 241–256.https://doi.org/10.1080/13642987.2019.1677623.

Keil, K., and Knecht, S., eds., 2017. *Governing Arctic change—Global perspectives*. Basingstoke, Hampshire: Palgrave Macmillan.

Khare, N., and Khare, R., 2021. *The Arctic: A barometer of global climate variability*. Amsterdam: Elsevier.

Knecht, S., and Keil, K., 2013. Arctic geopolitics revisited: Spatialising governance in the circumpolar North. *The Polar Journal*, 3 (1), 178–203. https://doi.org/10.1080/2154896X.2013.783276.

Kranholdt, D., 2022. Have we failed to capitalise on the momentum for the SDGs? (The Current Column 29.08.2022). Bonn: IDOS. Available from: https://www.idos-research.de/fileadmin/migratedNewsAssets/Files/German_Institute_of_Development_and_Sustainability_Kranholdt_29.08.2022.pdf [Accessed 25 November 2024]

Koivurova, T., *et al.*, 2022. Arctic cooperation in a new situation: Analysis on the impacts of the Russian war of aggression. Available from: https://julkaisut.valtioneuvosto.fi/handle/10024/164521 [Accessed 30 March 2023].

Larsen, J. N., and Fondahl, G., 2014. Arctic human development report. Regional processes and global linkages. Copenhagen: Nordic Council of Ministers. Available from: https://norden.diva-portal.org/smash/get/diva2:788965/FULLTEXT03.pdf [Accessed 5 July 2019].

Laruelle, M., 2019. The three waves of Arctic urbanisation. Drivers, evolutions, prospects. *Polar Record*, 55 (1), 1–12. https://doi.org/10.1017/S0032247419000081.

Lindberg, M., *et al.*, 2020. Co-creative place innovation in an arctic town. *Journal of Place Management and Development*, 13 (4), 447–463. https://doi.org/10.1108/JPMD-02–2019-0009.

Łuszczuk, M., *et al.*, 2022. Governability of regional challenges: The Arctic development paradox. *Politics and Governance*. 10 (3), 12. Available from: https://doi.org/10.17645/pag.v10i3.5341 [Accessed 12 February 2024].

Mathiesen, K., 2023. The green energy revolution's first casualties: Sweden's reindeer herders. *POLITICO*. Available from: https://www.politico.eu/article/ronnback-mine-sweden-reindeer-herders-sami-green-energy-revolution-first-casualties/ [Accessed 1 February 2024].

Mawdsley, E., 2021. Development Finance and the 2030 Goals. In: Chaturvedi, S., et al., eds., *The Palgrave Handbook of Development Cooperation for Achieving the 2030 Agenda*. Palgrave Macmillan, Cham. Available from: https://doi.org/10.1007/978-3-030-57938-8_3 [Accessed 24 November 2024].

Mayring, P., 2010. Qualitative Inhaltsanalyse. *In*: Mey, G., and Mruck, K., eds., *Handbuch Qualitative Forschung in der Psychologie*. Wiesbaden: VS Verlag für Sozialwissenschaften, 601–613.

McCauley, D., *et al*., 2022. Which states will lead a just transition for the Arctic? A DeePeR analysis of global data on Arctic states and formal observer states. *Global Environmental Change*, 73, 102480. https://doi.org/10.1016/j.gloenvcha.2022.102480.

Methi, K., and Wehrmann, D., 2023. Arctic cooperation beyond geopolitics. From Rovaniemi to Tromsø spirit? Available from: https://www.idos-research.de/uploads/media/German_Institute_of_Development_and_Sustainability_EN_Methi_Wehrmann_30.05.2023.pdf [Accessed 7 August 2023].

Metzger, J., and Lindblad, J., eds., 2020. *Dilemmas of sustainable urban development. A view from practice*. New York: Routledge.

Mills, G., 2007. Cities as agents of global change. *International Journal of Climatology*, 27, 1849–1857. https://doi.org/10.1002/joc.1604.

NATO Parliamentary Assembly 2017. "What happens in the Arctic, does not stay in the Arctic"—Climate change in the Arctic will have global consequences and cannot be ignored. [Press release]. Available from: https://www.nato-pa.int/news/what-happens-arctic-does-not-stay-arctic-climate-change-arctic-will-have-global-consequences [Accessed 16 May 2024].

Nordregio 2018. Municipalities by degree of urbanisation and functional urban areas. Available from: https://nordregio.org/maps/municipalities-by-degree-of-urbanisation-and-functional-urban-areas/ [Accessed 23 February 2023].

Normann, S., 2021. Green colonialism in the Nordic context: Exploring Southern Saami representations of wind energy development. *Journal of Community Psychology*, 49 (1), 77–94. https://doi.org/10.1002/jcop.22422.

Norwegian Human Rights Institution 2023. About the wind farms on Fosen and the Supreme Court judgment. Available from: https://www.nhri.no/en/2023/about-the-wind-farms-on-fosen-and-the-supreme-court-judgment/ [Accessed 30 January 2024].

Nyseth, T., and Pedersen, P., 2014. Urban Sámi identities in Scandinavia: Hybridities, ambivalences and cultural innovation. *Acta Borealia: A Nordic Journal of Circumpolar Societies*, 31 (2), 131–151. https://doi.org/10.1080/08003831.2014.967976.

Nystø Keskitalo, A. M., and Götze, J., 2023. Green transition and Indigenous Peoples' rights in the European Arctic. A rights-based approach for implementing the European Green Deal. *The Current Column*. Available from: https://www.idos-research.de/uploads/media/German_Institute_of_Development_and_Sustainability_EN_Keskitalo_Goetze_06.02.2023.pdf. [Accessed 12 March 2024].

Petrov, A. N., *et al*., 2017. *Arctic sustainability research: Past, present and future*. London: Routledge. Pram Gad, U., and Strandsbjerg, J., eds., 2019. *The politics of sustainability in the Arctic. Reconfiguring identity, space, and time*. New York: Routledge.

Pram Gad, U., Jacobsen, M., and Strandsbjerg, J., 2019. Introduction sustainability as a political concept in the Arctic. *In*: Pram Gad, U., Jacobsen, M., and Strandsbjerg, J., eds., *The politics of sustainability in the Arctic*. London: Routledge, 1–18.

Pries, L., 2010. *Transnationalisierung: Theorie und Empirie grenzüberschreitender Vergesellschaftung*. Wiesbaden: VS Verlag für Sozialwissenschaften.

Rantanen, M., *et al*., 2022. The Arctic has warmed nearly four times faster than the globe since 1979. *Communications Earth & Environment*, 3 (1), 168. https://doi.org/10.1038/s43247-022-00498-3.

Reinar, M. B., and Lundberg, A. K., 2023. Goals à la carte: Selective translation of the Sustainable development goals in strategic municipal planning in Norway. *Journal of*

Environmental Planning and Management, 67(11), 1–17. doi:10.1080/09640568.2023.2191816.

Rumjaun, A., and Narod, F., 2020. Social learning theory—Albert Bandura. *In:* Akpan, B., and Kennedy, T. J., eds., *Science education in theory and practice: An introductory Guide to learning theory*. Cham: Springer International Publishing, 85–99.

Sachs, J. D., *et al.*, 2019. Six transformations to achieve the sustainable development goals. *Nature Sustainability*, 2 (9), 805–814. https://doi.org/10.1038/s41893-019-0352–9.

Sejersen, F., 2019. Scaling sustainability in the Arctic. *In*: Pram Gad, U. and Strandsbjerg, J., eds., *The politics of sustainability in the Arctic. Reconfiguring identity, space, and time*. New York: Routledge, 94–107.

Simkiv, L., Shults, S., Lutskiv, O., and Andrusiv, U., 2021. Analysis of the dynamics of structural processes in the context of ensuring sustainable development. *European Journal of Sustainable Development*, 10 (1), 153–167. https://doi.org/10.14207/ejsd.

Spence, J., Alexander, E., Rødven, R., and Harriger, S., 2023. What makes the Arctic and its governance exceptional? Stories of geopolitics, environments and homelands. *Arctic Yearbook*. Available from: https://arcticyearbook.com/arctic-yearbook/2023/2023-scholarly-papers/499-what-makes-the-arctic-exceptional-stories-of-geopolitics-environments-and-homelands [Accessed 12 January 2024].

Steinberg, P., Tasch, J., & Gerhardt, H., 2015. *Contesting the Arctic: Politics and Imaginaries in the Circumpolar North*. I.B. Tauris

Stringer, T., Cheng, H. S., and Kim, A. M., 2023. A comparison of remoteness indices. *Polar Geography*, 46 (2–3), 95–119. https://doi.org/10.1080/1088937X.2023.2238792.

Sudarmadi, S., *et al.*, 2001. A survey of perception, knowledge, awareness, and attitude in regard to environmental problems in a sample of two different social groups in Jakarta, Indonesia. *Environment, Development and Sustainability*, 3 (2), 169–183. https://doi.org/10.1023/A:1011633729185.

Svennevig, I., 1997. Hvem er "de lokale" i naturbevaring? *Humanities Research Center: Man & Nature*, Working Paper 108. Available from: https://bibliotek.dk/materiale/hvem-er-de-lokale-i-naturbevaring-_ingeborg-svennevig/work-of:800010-katalog:99122016043205763?type=bog [Accessed 24 November 2024].

Tennberg, M., and Strauss-Mazzullo, H., 2023. Everyday practices of adaptation in the modern Arctic. *In*: Strauss-Mazzullo, H., and Tennberg, M., eds., *Living and working with snow, ice and seasons in the modern Arctic. Everyday perspectives*. Cham: Springer International Publishing, 1–13.

Thisted, K., and Gremaud, A.-S. N., 2020. *Denmark and the New North Atlantic. Narratives and memories in a former empire*. Aarhus, Denmark: Aarhus University Press.

Toy-Cronin, B., 2018. Ethical issues in insider-outsider research. *In*: Iphofen, R., and Tolich, M., eds., *The SAGE handbook of qualitative research ethics*. London: SAGE, 455–469.

UN 1987. Report of the World Commission on Environment and Development: note / by the Secretary-General. Available from: https://digitallibrary.un.org/record/139811?v=pdf [Accessed 24 November 2024].

UNEP 2005. Ecosystem and human well-being—synthesis. Available from: https://www.unep.org/resources/report/ecosystem-and-human-well-being-synthesis [Accessed 12 June 2024].

UN-Habitat 2024. New urban agenda. What we are doing. Available from: https://unhabitat.org/about-us/new-urban-agenda [Accessed 12 June 2024].

UN Sustainable Development Working Group 2024. Universal values. Principle two: Leave no one behind. Available from: https://unsdg.un.org/2030-agenda/universal-values/leave-no-one-behind [Accessed 12 June 2024].

United Nations 2017. New urban agenda. Available from: https://habitat3.org/wp-content/uploads/NUA-English.pdf [Accessed 3 April 2022].

United Nations 2021. UN chief promotes "enormous" benefits of greener cities. Available from: https://news.un.org/en/story/2021/10/1101992 [Accessed 14 August 2022].

United Nations 2023. Halfway to 2030, world "nowhere near" reaching global goals, UN warns. Available from: https://news.un.org/en/story/2023/07/1138777 [Accessed 1 December 2024].

Wall, C., and Wegge, N., 2023. The Russian Arctic threat. Consequences of the Ukraine war. Available from: https://csis-website-prod.s3.amazonaws.com/s3fs-public/2023-01/230125_Wall_RussianArcticThreat_0.pdf?VersionId=e8h73TdoOUjdJO3Y4nOTc4v5YRmpoZad [Accessed 9 February 2024].

Wehrmann, D., 2016a. The polar regions as "barometers" in the Anthropocene: Towards a new significance of non-state actors in international cooperation? *The Polar Journal*, 6 (2), 379–397. https://doi.org/10.1080/2154896X.2016.1241483.

Wehrmann, D., 2016b. Shaping changing circumpolar agendas: The identification and significance of "Emerging Issus" addressed in the Arctic Council. *The Arctic Yearbook 2016*, 90–103. Available from: https://arcticyearbook.com/images/Articles_2016/scholarly-articles/5-AY2016-Wehrmann.pdf [Accessed 18 April 2022].

Wehrmann, D., 2020. The Arctic Council as a success case for transnational cooperation in times of rapid global changes? *Arctic Yearbook*, 2020, 425–442. Available from: https://issuu.com/arcticportal/docs/ay2020?fr=sMzBiMzIyMjA2MjA [Accessed 5 September 2021].

Wehrmann, D., *et al.*, 2022. What future for cooperation in the Arctic? Scenarios after Putin's war on Ukraine. *Future of Globalisation* (blog). Available from: https://blogs.ids-research.de/2022/03/16/what-future-for-cooperation-in-the-arctic-scenarios-after-putins-war-on-ukraine/ [Accessed 14 July 2023].

Wendt, A., 1992. Anarchy is what states make of it: The social construction of power politics. *International Organization*, 46 (2), 391–425. https://doi.org/10.1017/S0020818300027764.

Wheeler, S. M., 2022. *The sustainable urban development reader*. London: Routledge.

Whitham, S., Sterling, L., Lin, C. E., and Wood, J. J., 2021. Social cognitive learning theory. *In:* Volkmar, F. R., ed., *Encyclopedia of autism spectrum disorders*. Cham: Springer International Publishing, 4418–4427.

2 Envisioning Sustainable Urban Development in Remote Regions

2.1 Introduction to Sustainable (Urban) Development

The previous chapter provided an overview of the challenges for pursuing sustainable urban development (SUD) in the European Arctic. This chapter contextualises the research perspective presented in this book by

- discussing global visions for advancing SUD and
- introducing the key components that feed into our conceptual model for analysing local approaches to SUD in remote regions.

This section focuses on how SUD is envisioned in global governance instruments and scholarly works. We proceed as follows. First, we discuss how global governance instruments intend the advancement of sustainable (urban) development. Here, we illustrate the extent to which multi-level and multi-actor approaches shall carry forward the implementation of the sustainable (urban) development goals. Second, we discuss the potentials and limitations of multi-level and multi-actor approaches by considering the scholarly literature on sustainable development.[1]

2.1.1 Visions in Global Governance Instruments

Three instruments with global coverage are central for our understanding of sustainable development and the relevance ascribed to urban places:

- The 2030 Agenda for Sustainable Development (adopted in 2015),
- The Paris Agreement (adopted in 2015, entered into force in 2016), and
- The New Urban Agenda (adopted in 2016).

These three instruments were negotiated under the auspices of different United Nations organisations, follow a universal aspiration and their goals are interlinked.[1] However, the Paris Agreement and the New Urban Agenda are directed at specific goals, whereas the 2030 Agenda provides a multi-dimensional understanding of sustainable development. Moreover, sustainable development is the central concept driving the 2030 Agenda and the New Urban Agenda but is "only" a reference point in the Paris Agreement.

DOI: 10.4324/9781003284864-2

Despite these differences, all three instruments prioritise two governance approaches:

1 governance across multiple levels and
2 an inclusive and participatory multi-actor approach.

Both approaches are inherent in the framing of the global goals and are considered crucial for their implementation. In the paragraphs that follow, we show how the three instruments relate to governance across multiple levels and an inclusive, participatory multi-actor approach in the framing of the goals and their envisioned implementation. While the different legal context of the three instruments and the national approaches to their implementation are elaborated in Chapter 3, it is important to note that the 2030 Agenda and the Paris Agreement do not differentiate between the geographical location of countries, nor do they address the factor of remoteness. The New Urban Agenda is the only instrument of the three that addresses "urban-rural interactions" and further recognises that especially urban centres located

> in developing countries, often have characteristics that make them and their inhabitants especially vulnerable to the adverse impacts of climate change [...], which particularly affect coastal areas, delta regions and small island developing States, among others.
>
> (UN 2017, p. 18)

The New Urban Agenda, however, neither draws specific attention to the Arctic regions nor to cities located in developed countries, even though, in contrast to cities in more southern regions, the specifics of Arctic cities, challenge more traditional understandings of urban spaces due to their remote context, differences in size, demographic structure, climate, and built environment (Nyseth 2017).

2.1.2 Multi-Level Governance (MLG) to Pursue Sustainable (Urban) Development

The goals and envisioned means of implementation, which the 2030 Agenda, the Paris Agreement, and the New Urban Agenda stress, are all shaped by and based on a MLG perspective. When considering the goals agreed upon, the Paris Agreement, for example, recognises under one of its three goals—mitigation, adaptation, and greening finance flows—"that adaptation is a global challenge faced by all with local, subnational, national, regional and international dimensions" (Article 7). Also, the parties to the New Urban Agenda "resolve to implement the New Urban Agenda as a key instrument for enabling national, subnational and local governments and all relevant stakeholders to achieve sustainable urban development" (UN 2017, p. 9). Similarly, with regard to the different policy fields that the concept of sustainable development pertains to, all instruments emphasise synergies across governance levels. The three instruments also envision a multi-level approach for the implementation of the objectives. They emphasise, for instance, the aim to "revitalize the global partnership for sustainable development" (SDG 17 2030 Agenda), to "foster stronger coordination and cooperation among national, subnational and local governments, including through multilevel consultation and mechanisms" (UN 2017, p. 23), and "[r]ecogniz[e] the importance of the engagements of all levels of government and various actors" (UNFCCC 2016, p. 3).

More specifically, the 2030 Agenda considers cooperation across governance levels crucial for financing the envisioned transformations (SDG 17.1) and for providing "access to science, technology and innovation" (SDG 17.6). The Paris Agreement (Article 7) also highlights the need for cooperation to enhance action and adaptation by sharing information, strengthening institutional arrangements and scientific knowledge, "[a]ssisting developing country Parties in identifying effective adaptation practices" and "[i]mproving the effectiveness and durability of adaptation actions" (UNFCCC 2016, p. 10). The New Urban Agenda emphasises the vertical and horizontal dimensions of MLG when relating to the "coordination role of national, subnational and local governments [...] and their collaboration with other public entities and non-governmental organizations" by focusing specifically on "the provision of social and basic services" (UN 2017, p. 12). It further flags the need to "enabl[e] policy frameworks at the national, subnational and local levels, integrated by participatory planning [...]" (UN 2017, p. 22). While this focus on participation and partnerships also corresponds to the Global Partnership approach brought forward in the 2030 Agenda (which aims at "bringing together Governments, civil society, the private sector, the United Nations system and other actors", Article 60), the New Urban Agenda adds the specific aspect of planning.

Further, both the 2030 Agenda and the New Urban Agenda demand coordination. Coordination is considered essential for policy alignment, on the one hand, and to bring different stakeholders actors together, on the other. In regard to policy alignment, the 2030 Agenda aims at enhancing policy coherence for sustainable development (SDG 17.14) and the parties "commit to pursuing policy coherence and an enabling environment for sustainable development at all levels and by all actors, and to reinvigorating the global partnership for sustainable development". The New Urban Agenda also considers "sustainable urban and territorial development as part of integrated development strategies and plans, supported, as appropriate by national, subnational and local institutional and regulatory frameworks" (UN 2017, p. 23). With regard to the horizontal dimension of MLG, both instruments "promote effective public, public-private and civil society partnerships" (SDG 17.17) and

> invite international and regional organizations and bodies, including those of the United Nations system and multilateral environmental agreements, development partners, international and multilateral financial institutions, regional development banks, the private sector and other stakeholders, to enhance coordination of their urban and rural development strategies and programmes to apply an integrated approach to sustainable urbanization, mainstreaming the implementation of the New Urban Agenda.
>
> (UN 2017, p. 22)

Table 2.1 Framing of MLG in global instruments for sustainable (urban) development

	The 2030 agenda for sustainable development	*The Paris agreement*	*The new urban agenda*
Goals of cooperation:	Cooperation to finance transformations, to provide access to science, technology, and innovation (SDG 17.1).	Cooperation for enhancing action and adaptation by sharing information, strengthening institutional arrangements, and scientific knowledge (Article 7).	Collaboration with other public entities and non-governmental organisations to provide social and basic services and enable participatory planning.
Requirements for implementation: coordination, integration, and participation	Coordination to enhance policy coherence and enable sustainable development at all levels and by all actors (SDG 17.14).	Need to enhance public and private sector participation and coordination across instruments and institutions.	Integrated development strategies and plans supported by institutional and regulatory frameworks.

Own work.

Both instruments thereby build on the understanding that multi-actor partnerships are needed to achieve the goals.[2] Likewise, Article 6 in the Paris Agreement recognises in the context of market mechanisms that the emissions of greenhouse gas and sustainable development are interlinked as well as the need to enhance public and private sector participation and coordination across instruments and institutions. As summarised in Table 2.1, the instruments under analysis promote MLG to provide access to finances, knowledge, social and basic services and to demand coordination, integration, and participation, in particular, to pursue these goals across governance levels. The following section provides examples on how the global governance instruments frame such an inclusive, participatory multi-actor approach in their goals and envisioned implementation strategies.

2.1.3 An Inclusive and Participatory Multi-Actor Approach to Pursue Sustainable (Urban) Development with and for All

The goals of the 2030 Agenda, the Paris Agreement, and the New Urban Agenda follow a global aspiration. To achieve "sustainable and inclusive urban prosperity and opportunities for all" (UN 2017, p. 14), all instruments accordingly stress an inclusive and participatory approach, which is the basis for the goals agreed upon and envisioned for their implementation.[3] While both the Paris Agreement and the 2030 Agenda address particularly the country level, they do not emphasise the differences *within* countries, which are of great relevance when considering regional diversities.[4] By contrast, the New Urban Agenda is the only instrument that emphasises "the key role of cities and human settlements as drivers of sustainable development in an increasingly urbanized world" (UN 2017, p. 11). This further relates particularly to the dimension of local governance (different to the Paris Agreement and the 2030 Agenda):

1 "We support subnational and local governments, as appropriate, in fulfilling their key role in strengthening the interface among all relevant stakeholders, offering opportunities for dialogue" (UN 2017, p. 14) and
2 "We encourage effective participation and collaboration among all relevant stakeholders, including local governments, the private sector and civil society, women, organizations representing youth, as well as those representing persons with disabilities, indigenous peoples, professionals, academic institutions, trade unions, employers' organizations, migrant associations and cultural associations, in order to identify opportunities for urban economic development and identify and address existing and emerging challenges" (UN 2017, p. 15).

The 2030 Agenda, on the other hand, refers only to the regional and sub-regional dimensions to "facilitate the effective translation of sustainable development policies into concrete actions at national level" (Article 21).

Concerning the implementation of the goals agreed upon, the inclusive partnership-approach brought forward in the 2030 Agenda is based on "a spirit of global solidarity, in particular with the poorest and with people in vulnerable situations" and encompasses "Governments, the private sector, civil society, the United Nations system and other actors" (Article 39). The New Urban Agenda more specifically seeks to "engage indigenous peoples and local communities in the promotion and dissemination of knowledge" (UN 2017, p. 32) and "support[s] strengthening the capacity of subnational and local governments to implement effective local and metropolitan multilevel governance, across administrative borders […] ensuring the involvement of subnational and local governments in decision-making" (UN 2017, p. 23). All three instruments thereby follow a context-sensitive approach and consider different capacities of the actors involved in the implementation of the global goals. In this regard, the 2030 Agenda and the New Urban Agenda stress the primary responsibility of each country "for its own economic and social development" (Article 41 and UN 2017, p. 33). The 2030 Agenda further acknowledges that "each government [is] setting its own national targets guided by the global level of ambition but taking into account national circumstances" (Article 55). In its overarching goals, the Paris Agreement (UNFCCC 2016, p. 4) emphasises the need, "to reflect equity […] in the light of different national circumstances".

To realise an inclusive, participatory approach for implementing the global goals, all agreements focus on capacity development and capacity building. The 2030 Agenda demands "targeted capacity-building" particularly in developing countries (SDG 17.9). Also, the Paris Agreement (Article 11) stresses the need to enhance "capacity and ability of developing country Parties, in particular countries with the least capacity, such as the least developed countries, and those that are particularly vulnerable to the adverse effects of climate change, such as small island developing States". The New Urban Agenda, again, is the only instrument taking a local and pluralistic stance to actor engagement, when promoting "capacity development as a multifaceted approach that addresses the ability of multiple stakeholders and institutions at all levels of governance" (UN 2017, p. 37). In regard to capacity building, the New Urban Agenda further paves the ground for a transnational approach by stressing the aim to expand opportunities for "subnational, decentralized and city-to-city cooperation" (UN 2017, p. 36)[5] and to provide "support to local governments in partnering with communities, civil society and the private sector" (p. 24) without limiting these to the country level.

Table 2.2 summarises how the global governance instruments under analysis frame participation to develop effective policies that acknowledge context sensitivities, different capabilities, and the principle of shared responsibility. To pursue this goal, they all call for capacity building.

The three global instruments further apply different approaches to hold different actors accountable for their actions. In this regard, the New Urban Agenda emphasises its

> support of local governments and relevant stakeholders, through a variety of mechanisms, in developing and using basic land inventory information, such as cadastres, valuation and risk maps, and land and housing price records, to generate the high-quality, timely and reliable data […] needed to assess changes in land values, while ensuring that these data will not be used for discriminatory land-use policies.
>
> (UN 2017, p. 27)

In contrast, the 2030 Agenda and the Paris Agreement address and stress the role of national governments.[6] The countries adopting or ratifying these instruments commit themselves to contribute to systematic follow-up and review processes to "promote accountability to our citizens, support effective international cooperation in achieving this Agenda and foster exchanges of best practices and mutual learning" (2030 Agenda, Article 73) and participate in an enhanced transparency framework "in order to build mutual trust and confidence and to promote effective implementation" (Article 13, Paris Agreement). Moreover, the 2030 Agenda "encourages member states to conduct regular and inclusive reviews of progress at the national and sub-national levels which are country-led and country-driven" (2030 Agenda, Article 79).

2.1.4 Governance Instruments to Advance Sustainable (Urban) Development

MLG and multi-actor approaches are deep-seated in the framing of the 2030 Agenda for Sustainable Development, the Paris Agreement, and the New Urban Agenda. Both MLG and inclusive, participatory multi-actor approaches further interlink in their functions. The governance instruments describe MLG and multi-actor approaches as enabling governments and actors across policy-fields and governance levels to

Table 2.2 Framing of participation in global governance instruments for sustainable (urban) development

	The 2030 agenda for sustainable development	*The Paris agreement*	*The new urban agenda*
Goals of participation: Acknowledgement of context-sensitivity and shared responsibility	To facilitate the translation of sustainable development policies into concrete actions at national level (Article 21). To foster inter-cultural understanding, tolerance, mutual respect, ethic of global citizenship, shared responsibility of all to contribute and enable sustainable development (Article 36).	To reflect equity and the principle common but differentiated responsibilities and respective capabilities, in the light of different national circumstances.	To develop local, durable, and dignified solutions.
Requirements for implementation: Capacity building	Targeted capacity-building particularly in developing countries (SDG 17.9).	Enhancing the capacity and ability, in particular of developing countries and those that are particularly vulnerable to the adverse effects of climate change.	Institutional, political, legal, and financial mechanisms to broaden inclusive platforms for participation in decision-making, planning and follow-up processes, for enhanced civil engagement and co-provision and co-production. Strengthening the capacity of subnational and local governments for ensuring the involvement of subnational and local governments in decision-making.

Own work.

pursue the goals agreed upon, among others, by taking advantage of synergies, and providing access to information and finances. To fulfil these functions, the three instruments emphasise the importance of collaboration across different levels of governance and stress the necessity of bolstering institutional frameworks. Specifically, they highlight the critical roles of national, subnational, and local governments, their partnerships with other public bodies and non-governmental organisations and the need to foster coordination and cooperation to facilitate engagement and partnerships at all levels by strengthening institutional arrangements and participation.

The agreements, however, do not discuss challenges that arise from pursuing both approaches. Even though they acknowledge context-sensitivities such as differences in capacities at the country level and in state and non-state actor engagement, the question of how power differentials may shape coordination and cooperation across governance levels and among various (unequal) actors remains unaddressed. How these context-sensitivities, which are also reflected in principles such as "common but differentiated responsibilities and respective capabilities, in the light of different national circumstances" (UNFCCC 2016), are addressed in MLG and multi-actor approaches is thus left to further (political) negotiations.

2.2 Governance Approaches for Sustainable Development in Scholarly Works

In this section, we explore the potential and limitations of MLG structures and multi-actor approaches for advancing sustainable (urban) development by bringing together different strands of academic literature. First, we shed light on how the operationalisation of both approaches is discussed in global and local governance studies by considering the concepts of polycentric governance and indirect governance. Second, we focus specifically on how MLG and multi-actor approaches have been reviewed in the context of sustainable (urban) development, by bringing together literature published in the fields of sustainability studies and urban studies.

2.2.1 MLG Structures and Multi-Actor Approaches in Governance Studies

Irrespective of the governance level they pertain to, political orders are social constructs that can be understood as

> assemblages of institutional arrangements either created intentionally by human actors seeking to address some consciously delimited domain of human affairs or evolving through recurrent social interactions as distinct and generally stable practices dealing with more-or-less well-defined spheres of human affairs.
>
> (Young 2023, p. 2)

Multi-actor partnerships[7] operate within these political orders (= the operating environment), which are dynamic and shaped by "shifts in the capabilities, preferences, and practices of human actors responding to a variety of biophysical and

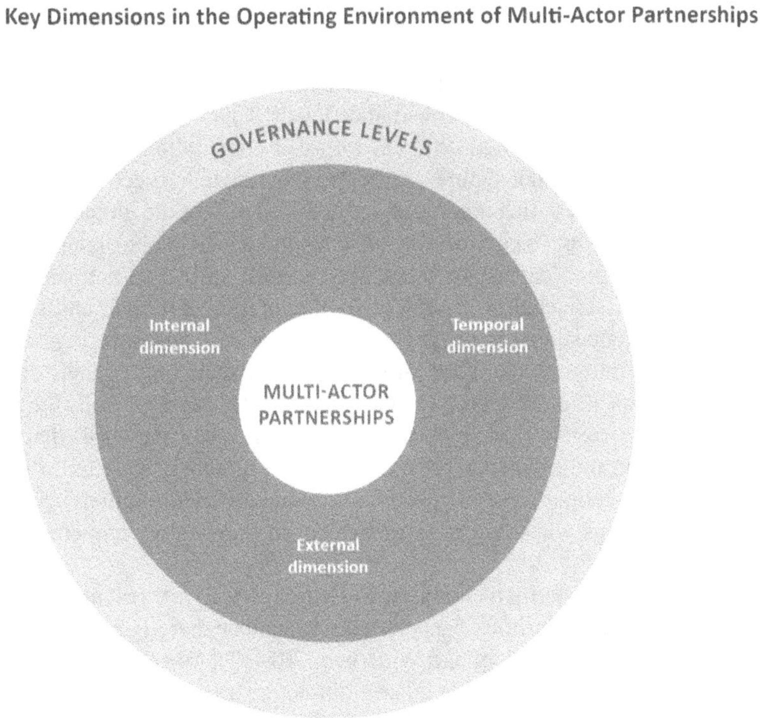

Figure 2.1 Key dimensions in the operating environment of multi-actor partnerships © IDOS.

socioeconomic developments" (Young 2023, p. 2). Multi-actor partnerships thus take into account an external dimension, such as the evolution of political orders and the related institutional arrangements and social interactions, and an internal dimension, with dynamics driven by the human actors involved.

In addition to the external and internal dimensions that encourage change, the political order in all countries with territory in the European Arctic is based on the principle of democracy, with policies and decisions determined among other types of decisions by electoral cycles. Transformations in policy- and decision-making (the structures, themes, and the actors involved) thus change "naturally" over the course of time (temporal dimension) but are also affected by changing conditions beyond their national jurisdictions, as most recently with the COVID-19 pandemic and Russia's full-scale invasion of Ukraine (Figure 2.1).

Also, against this backdrop, research in the disciplines of International Relations (IR), global and environmental governance, and development studies has convincingly stressed the need to consider global complexities and local sensitivities when examining the interconnected global goals and targets (Barber and Bartlett 2021, Chaturvedi *et al.* 2021, Horner and Hulme 2017). In this vein, research has focused on the transformation of the global governance architecture (e.g., the Earth System

Governance Project), the future trajectory of the global political order, and the "growing collection of unprecedented needs for governance" with which the world is confronted (Young 2021). Scholars have considered the diversity of partnerships operating at and across different governance levels (Chan *et al.* 2018, McAllister and Taylor 2015) and identified different forms of cooperation and imaginations that "guide cooperative endeavours so that they take certain directions" (Freistein *et al.* 2022, p. 1), all of which are important in the pursuit of the global goals.

The global goals are based on the understanding that global cooperation is needed to address problems that cannot be solved by unilateral actions (Adams and Martens 2015, Treichel *et al.* 2016). As the internal, external, and temporal dimensions mentioned above suggest, the potential for social engineering is limited in terms of cooperation guidance, as "it is hard to anticipate how specific arrangements will operate in a complex system" (Young 2023, p. 56). Still, the global agreements create certain principles and expectations (see Chapter 4) that actors who collaborate in multi-actor partnerships commit to adhere to (e.g., to operate on equal bases in longer-term partnerships to pursue transformations). But how do these partnerships work and to what extend do they follow/implement these principles in practice?

Ideally, multi-actor partnerships form to share knowledges, resources, competences, risks, and responsibilities that are needed to achieve shared goals (cf. SDG 17, Bäckstrand *et al.* 2012, Pattberg and Widerberg 2014, Thorpe and Maestre 2015). In this way, researchers also conceptualise multi-actor partnerships often in positive terms, for example, as "mechanisms to help resolve a variety of current governance deficits" and solutions-oriented "innovative arrangements" (Pattberg and Widerberg 2014, p. 9 and Young 2023, p. 53). Given the "growing influence of various types of non-state actors" including multi-actor partnerships, some even presume that in a future global political order states will not remain "the sole repositories of political authority" (Young 2023, p. 55). As the dimensions introduced above suggest, multi-actor partnerships do not exist in isolation. They and the actors involved compete for resources and influence and are embedded in different contexts that require a close analysis of their specificities (Chan 2016, Wehrmann 2018). Moreover, the actors involved in multi-actor partnerships have different "priorities, values and attributes" (Tennyson 2011, p. 5), and their cooperation is shaped by power relations that are determined not only by political orders (external dimension) but also by individual capacities and capabilities (internal and temporal dimensions).

Falling short of the ideal, progress towards achieving the SDGs has been limited (United Nations 2023). The experience of polycrisis (the interconnection of global crises such as the COVID-19 pandemic, the climate crisis, and the crisis of multilateralism) and contestation in global governance (Chaturvedi *et al.* 2021) has contributed to a "reversal in progress" (Kranholdt 2022). However, these crises only partly explain the foreseeable failure to achieve the global goals and targets: the Global Partnership-approach promoted by SDG 17 also fell short of expectations (Cruz 2023). The question of how to advance cooperation and the effectiveness of multi-actor partnerships seems particularly crucial in this regard (Andonova *et al.* 2022). Development researchers have published extensively on this issue and have evaluated a range of interventions (among the many, see Beisheim and Simon 2017, Loveridge and Wilson 2017, Prescott and Stibbe 2017). Most of them highlighted the need for leadership, such as a meta-governance for multi-actor partnerships providing coordination, incentives for cooperation, monitoring and regulation (Beisheim and Simon 2017, Wehrmann 2018). But considering the diversity of political orders, actors, relationships, and formats of cooperation across governance levels, what should such leadership ideally look like? And when should different forms of leadership be in place?[8]

In general terms, governance can be defined as "an inclusive system of actors, institutions and norms that establishes responsibility and accountability, and builds trust and capacity to cooperate in policymaking, decision-making, implementation and enforcement" (Nesbit *et al.* 2019). Such definition, however, does not relate to the quality of cooperation nor to leadership in governance, and both can take different forms (Fukuyama 2013, Léautier 2014). Given our focus on multi-actor approaches across governance levels, in what follows, we introduce two prominent concepts—*polycentric governance* and *indirect governance*—to discuss how leadership can be operationalised to promote coordination within and among multi-actor partnerships and the adherence to principles to pursue the global goals and targets at all levels of governance.

2.2.2 Polycentric Governance

The concept of polycentricity is based on the aim of advancing the management of the commons (Ostrom 2009, Rakhyun 2020), which is also central to the Paris Agreement and addressed in the 2030 Agenda for Sustainable Development. Vincent and Elinor Ostrom and their collaborators conceived of polycentric governance accordingly as a multilevel phenomenon (Liefferink and Wurzel 2018). Polycentric governance theory, however, is grounded on the premise that different authorities (meaning governing initiatives) operate in a layered landscape without standing in a hierarchical relationship to another (Morrison *et al.* 2017, Ostrom 2009, Pattberg *et al.* 2018). Yet, hierarchical governors of (national) societies, such as states and non-state actors alike, can form a part of polycentric governance. Particularly in global climate governance, scholars have provided evidence of how non-state, sub-national, and private actors as well as independent initiatives and climate actions form a networked—polycentric—structure (Pattberg *et al.* 2018).[9]

Within this networked structure, actors do not operate in isolation; they are able to collaborate and interlink. The linkages among the different actors can be created by all, however, their quality differs. Pattberg *et al.* (2018) differentiate between four types of institutional linkages:

1 Cognitive linkages (CL) that are grounded on the exchange of knowledge(s) and learning,
2 Linkages through commitment (LtC) that are based on the diffusion of norms and principles,

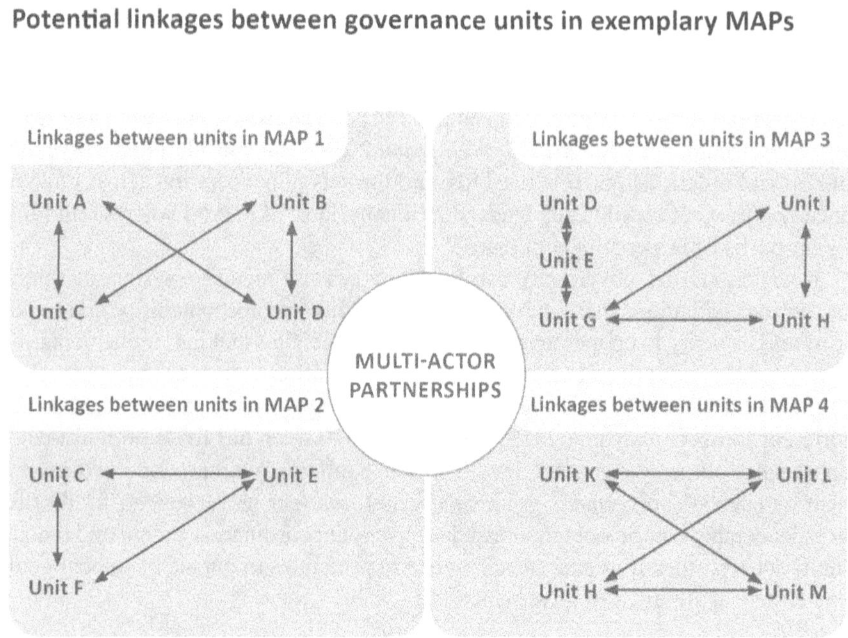

Figure 2.2 Potential linkages between governance units in exemplary MAPs. © IDOS.

3 Behavioural linkages (BL) that trigger behavioural changes, and
4 Impact-level linkages (I-lL) that affect how activities are pursued (Figure 2.2).[10]

Due to these linkages, which in global climate governance appear most often as synergistic instead of conflictive (Pattberg *et al.* 2018), actors are able to adjust. The polycentric governance structure thus "encourages the consideration of multiple perspectives and […] of local-specific conditions" (Morrison *et al.* 2017, p. 2). This adjustment is crucial as "individual behaviour is strongly affected by the context in which interactions take place" (Ostrom 2009, p. 431), and, likewise to pursue global agreements, "it is necessary to link them more strongly to individual contexts" (Jakobeit *et al.* 2010 cited Wehrmann 2020, p. 6).

While global agreements, such as the Paris Agreement, encourage and stimulate the linkages between state and non-state actors (Liefferink and Wurzel 2018), leadership is not provided by state actors per se. Instead, in a polycentric governance system (PGS), different actors take responsibility to coordinate interactions, communicate, or advance linkages between and among governance actors such as states, non-state actors, and multi-actor partnerships.[11] While any actor can become a leader or pioneer,[12] some have more economic power (e.g., large corporations) or social power (e.g., civil society organisations) than others.[13] Moreover, in polycentric governance, both leadership and pioneership "cannot be understood without taking into account their embeddedness in more hierarchical, top-down […] arrangements", and the effects of leaders and pioneers can be limited "to the relatively independent unit in which they function" (Liefferink and Wurzel 2018). With that in mind, Liefferink and Wurzel (2018) distinguish between four types of leadership in polycentric governance:

1 Structural leadership (through military and economic power, or an actor's systemic relevance),
2 Entrepreneurial leadership (via diplomatic, negotiating and bargaining skills—including the paradiplomatic activities of cities),
3 Cognitive leadership (based on knowledge and expertise), and
4 Exemplary leadership (intended/unintended provision of examples for other actors).

National governments, for example, may provide structural leadership by establishing platforms and coordinating multi-actor partnerships to pursue the global goals. In such multi-actor partnerships, cities, representatives of civil society organisations or businesses and researchers may provide entrepreneurial or cognitive leadership while—at the same time—the multi-actor partnership may provide exemplary leadership to other multi-actor partnerships, operating, for example, at another governance level, in another region, or towards different objectives. In this way, MLG researchers have paid particular attention to the subnational level and also to urban areas, which "are seen as 'polycentric' governance arrangements (e.g., Hall and Pain 2006) or as 'functional overlapping and competing jurisdictions' (Eichenberger and Frey 2006) that play an important role within a multi-level context (see Sellers 2002)" (Zürn *et al.* 2010, p. 7). More generally, scholars have identified different modes of leadership at several governance levels (Benulic *et al.* 2022).

The number and intensity of linkages and the number and engagement of actors involved obviously shape the polycentric structure of the governance system, but it is important to explore how conflicts and overlaps affect PGSs in general and also in specific cases. In this regard, research on the different stages of a PGS (e.g., whether it is emerging, stable, or persistent) as well as on its quality (e.g., network approaches) is emerging.[14] While polycentric governance has been criticised for leading to "high transaction costs, inconsistencies, freeloading, unanticipated effects, gridlock, and ultimate implementation failure" (Morrison *et al.* 2017, p. 2), PGSs are also perceived as more robust than others because significant overlaps and redundancy allow for replacement if parts of the system fail (Carlisle and Gruby 2019 cited Morrison *et al.* 2023). Moreover, polycentric governance "allows specialization, division of tasks between central, regional, and local levels, subsidiarity, and tailoring of interventions to local-regional circumstances and community preferences" and therefore improves efficiency "by matching interventions to the context and scale of the problem" (Morrison *et al.* 2017 cited Morrison *et al.* 2023) (Figure 2.3).[15]

In summary, in PGSs, the collaboration of different governing units (such as multi-actor partnerships, countries, regional forums, and transnational alliances) is characterised by different types of linkages and different forms of leadership and,

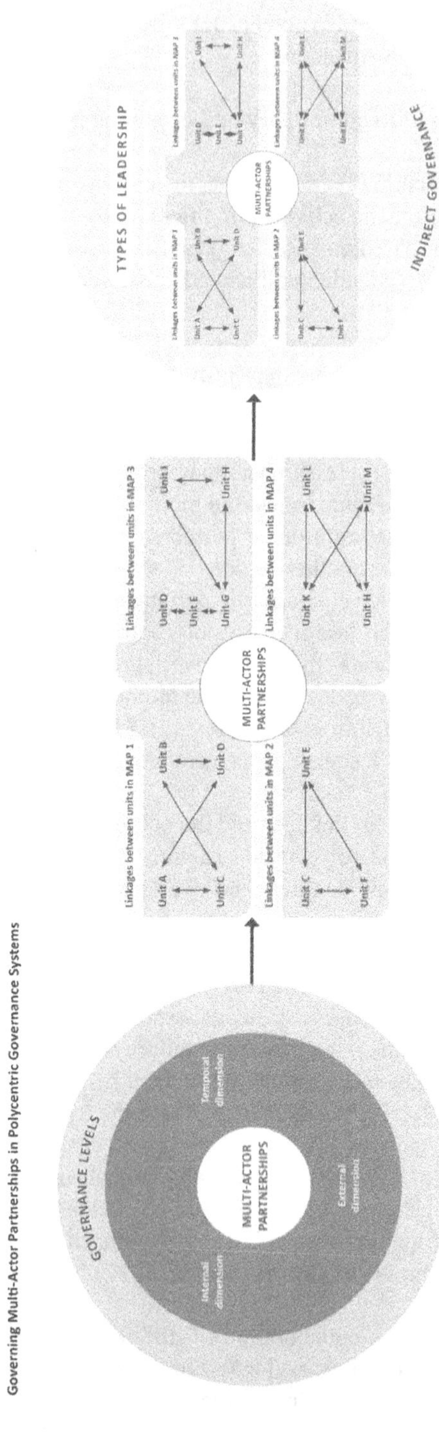

Figure 2.3 Governing multi-actor partnerships in PGSs © IDOS.

as in most governance, mostly steered by indirect governance (Abbott et al. 2018). To better understand how the actors involved can address collective challenges effectively through improved coordination[16] (in view of linkages and leadership), we shed light on two contrasting logics of indirect governance, namely delegation and orchestration.

2.2.3 Indirect Governance: Delegation

Governance is most often *indirect* as governors usually do not have sufficient "capabilities to govern single-handedly" (Abbott *et al.* 2018, p. 4). As principals, they bring in agents (such as other states, non-state actors, and international organisations) to enhance their effectiveness, efficiency, or legitimacy. Depending on the setting, authority either lies with the principal (in hierarchical settings) or with the agent (in non-hierarchical settings). In what follows, we explain two "pure modes" of indirect governance that illustrate both types of relationship:

- delegation, when an agent carries out governance tasks on behalf of the principal, and
- orchestration, when an agent is mobilised on a voluntary basis to pursue governance goals shared with a principal.[17]

Delegation models typically build on principal-agent theory, game theory, and new institutionalism in rational choice theory. Delegation is a (hierarchical) form of cooperation, which takes place if "the *principal* delegates authority to a specialized *agent* with the expertise, time, political ability, or resources to perform a task" (Hawkins *et al.* 2006, p. 13, emphasis added). Delegation is thus a "hard" and conditional form of governance "because the governor has formal legal control over the agent […] supervises its activities and can ultimately rescind its authority" (Abbott *et al.* 2021, p. 150).

Delegation can be applied in governance processes at and across different levels (Epstein and O'Halloran 1999, Pollack 1997) and is motivated among others by problems of coordination (Coen and Thatcher 2008). While delegation is not a prerequisite for international cooperation, it is "premised upon the division of labor and gains from specialization" (Hawkins *et al.* 2006, p. 13). Delegating powers may follow different logics of delegation, and, in this way, scholars provided evidence that "the nature of delegation widely differs across countries" with respective implications also on subnational governance (Enderlein 2010, p. 427).[18] Overall, however, the causes and consequences of delegation appear similar in international and domestic politics (Hawkins *et al.* 2006, p. 4).

The hierarchical arrangements in which delegation is most prevalent are found in national jurisdictions, which multilevel governance researchers refer to as Type I governance (e.g., Hooghe and Marks 2010). Type I governance is often rooted in national, regional, and/or local identity and "oriented to intrinsic communities and to their demands for self-rule" (Hooghe and Marks 2010, p. 28), while Type II jurisdictions are instrumental arrangements that "solve ad hoc coordination problems among individuals sharing the same geographical or functional space" (Hooghe and Marks 2010, p. 27). Transnational public-private partnerships[19] and multi-actor partnerships are found in both Type I and Type II governance but are more common in Type II and even more so beyond the national level. However, Type II governance may also appear at the local level when communities have "to cope with locally specific common pool resource problems (Ostrom 1990)" (Hooghe and Marks 2010, p. 26), for example, to restrict access to scarce renewable resources. In terms of *leadership*, when delegation is used, principals gain hierarchical control but also greater responsibility in relation to the agent, as the principal can and must monitor the agent in order to impose sanctions or withdraw support (e.g., resources) if the agent does not "perform in the way envisioned" (Hawkins *et al.* 2006, p. 8). This greater responsibility, however, limits the local legitimacy of both the agent and the principal and support for them (Heinkelmann-Wild and Mehrl 2022).

The conditions under which governors decide to apply different modes of indirect governance (such as delegation and orchestration) for *coordination* vary. The (in)ability to exercise hard controls, path-dependency and (missing) goal alignment seems to influence that choice, but the number of actors involved who compete for resources also matters (cf. Abbott *et al.* 2020, Heinkelmann-Wild and Mehrl 2022). Orchestration seems more likely to be applied in cases where efficiency losses and non-compliant behaviour can be afforded (Heinkelmann-Wild and Mehrl 2022). Both, delegation and orchestration, however, are "stark ideal types" that are usually mixed. To better understand how leadership in multi-actor partnerships can be operationalised to pursue the global goals and targets, we will focus on orchestration next.

2.2.4 Indirect Governance: Orchestration

In the literature on international regime and transnational governance complexes, orchestration is an indirect and soft mode of governance. Orchestration is understood "as the facilitation and coordination of intermediary actors on a voluntary basis by providing them material and ideational support in order to achieve governance goals with respect to target actors" (Abbott *et al.* 2012, p. 6). Different to traditional, hierarchical modes of governance (such as delegation), *leadership* is thus divided among orchestrators (one actor or a set of actors) and intermediaries (one actor or a set of actors) who govern the target (one actor or a set of actors). If there is no intermediary, and two actors (or group of actors) work together for a mutual goal, the mode of governance is not orchestration but collaboration (Chaturvedi *et al.* 2021, p. 16).

Orchestration, similar to delegation, is based on a pluralist view of global governance and can take place at and across different governance levels (Abbott *et al.* 2021). However, Abbott *et al.* (2016) argue that "orchestration is relatively more likely in democratic than authoritarian systems". In domestic orchestration, for example, a state may play the role of an orchestrator when it asks the municipalities (intermediaries) to pursue the global goals by setting specific guidelines for urban development that businesses (targets) must follow. International organisations or intergovernmental fora, such as the Arctic Council, can also be understood as orchestrators, for example, when they work with the Arctic states (as intermediaries) to implement agreements who then provide leadership to the subjects (targets) to whom the agreements pertain and who relate to the agreements in their actions accordingly.[20] In all cases, the orchestrating actor (or set of actors) "has no hard control over the activities of intermediaries but must mobilize and facilitate their voluntary cooperation in a joint governance effort" (Abbott *et al.* 2012, p. 6). However, as a mode of governance, orchestration itself is also a multi-actor system, based on the central assumption that multi-actor partnerships are formed (orchestrators, intermediaries, and targets cooperate) because the actors alone cannot achieve the common goals. The concept thus parallels the universalist framework of the SDGs. At the same time, the concept of orchestration acknowledges that the actors involved in multi-stakeholder partnerships "operate in an institutional context" and may also be shaped by different modes of governance (Abbott *et al.* 2021, p. 150).

While the literature on effectiveness in development studies has reviewed the plurality of actors with overlapping mandates and increasing fragmentation[21] more critically (Klingebiel *et al.* 2016), orchestration theory sees the co-existence of multiple governance actors with overlapping mandates as beneficial. In this way, it is argued that member organisations can be empowered through the "gains from specialization, pooling of resources and mutual learning" in organisational complexes (Abbott *et al.* 2021, p. 145). Moreover, multiple orchestrators could reinforce mutual efforts in different governance contexts and thus improve the overall quality of governance (Chan *et al.* 2018). By recognising differences in status and leadership, orchestration theory emphasises linkages (relationships and interactions) between governance actors and the *coordination* of (also catalytic) linkages.[22] But how are techniques of orchestration used in such interactions? With reference to the policy cycle, Abbott *et al.* (2021) show how actors can apply orchestration by (1) steering and empowering of actors for a specific purpose, (2) mobilising intermediaries and shaping goals, (3 and 4) endorsing the intermediary, and (5) through coordination. While the mix of techniques may vary, in view of leadership, orchestrators can thus

convene (1) with the actors they select and consider important and whom they bring together with other influential actors,
set the agenda (2) by defining or framing governance issues and potential policies,
provide assistance (3) to intermediaries via material support and (4) recognise intermediaries as competent and legitimate, and
synchronise activities (5) to increase the impact of intermediaries.

To sum up: In light of the question explored in this book—why the global goals are pursued differently in the European Arctic—this section sheds light on MLG structures and multi-actor approaches in governance research and introduces the characteristics of collaboration between different governing units across governance structures and in different multi-actor settings. We show that multi-actor partnerships operate within dynamic political orders. They change in response to at least three dimensions (internal, external, and temporal) and across levels of governance. Multi-actor partnerships are diverse, as are the ways in which they work together and the ways in which the actors involved imagine their cooperation themselves. The global goals are based on the understanding that multi-actor partnerships form to address shared challenges. Research on polycentric and network governance illustrated, however, that in a PGS, the cooperation of different kinds of actors is based on different types of linkages (such as cognitive linkages, linkage through commitment, behavioural linkages, and impact-level linkages). In addition to the dynamic political orders and the governance levels at which multi-actor partnerships operate, these linkages shape the quality of their cooperation. In these arrangements, leadership is not provided by state-actors per se. Instead, in multi-actor partnerships that operate in a PGS, different actors take responsibility to coordinate interactions, communicate, or advance linkages between and among governance actors such as states, non-state actors, and other multi-actor partnerships. Accordingly, leadership may take different forms, such as structural leadership, entrepreneurial leadership, cognitive leadership, and exemplary leadership. In these complex settings, however, collaboration is not steered by one actor/leader/pioneer alone. Instead, governing actors typically use indirect governance to address collective challenges, where delegation and orchestration are contrasting examples. Both modes of indirect governance can be applied at various governance levels and, often, they are mixed. Delegation as a hard mode of governance is more prevalent in hierarchical arrangements, while orchestration as a soft mode of governance is more likely in democratic systems. Orchestration parallels the universal approach of the 2030 Agenda and the Paris Agreement but delegation is also based on a pluralistic view of global governance. In terms of coordination, governing actors who use delegation have greater control and responsibility but may also face legitimacy gaps. Orchestrators rely on voluntary cooperation with intermediaries and targets. For them, linkages (relationships and interactions) are central to coordination and capacity building.

Building on this section, which examined MLG structures and multi-actor approaches in governance research to identify prerequisites and modes of collaboration between different governing units across governance structures and in different multi-actor settings, the following section focuses specifically on how MLG and multi-actor approaches have been reviewed in the context of sustainable (urban) development. As we show, different strands of research on sustainable (urban) development underpin the leadership and coordination approaches discussed earlier and provide further insights into the strategic choices that actors need to consider in multi-level and multi-actor arrangements for advancing SUD.

In what follows, we will provide an exemplary overview of these insights by focusing specifically on

- the special relevance ascribed to the local level for pursuing global goals in the Smart City discourse (1),
- the organisational challenges to coordinate multiple actors and advance policy-transfer (2),

- the role of the institutional set-up and leadership in multi-actor partnerships (3), and
- MLG and multi-actor approaches in polycentric and network perspectives (4).

2.3 Multilevel Governance and Multi-Actor Approaches in Sustainability Studies and Urban Studies

In relation to SUD, the Smart City discourse in particular focuses on multi-level governance and multi-actor partnerships[23] and acknowledges that "urban areas are at the forefront of global change", and their future development is crucial for the pursuit of the global goals (Mora *et al.* 2023, p. 1560). In this sense, smart cities and the *local level of government* (1) have to address multiple priorities (Britton 2019). To foster "local ownership" (Bilsky *et al.* 2021), research on the SDG localisation movement has emphasised the involvement of a plurality of actors and citizen participation in consultative approaches, in order to recognise the diversity of capacities in "ambitious multilevel governance arrangements and multistakeholder co-creation efforts" (Bilsky *et al.* 2021, p. 713). Here, the local level of government is further seen as "best placed […] to leverage their collective capabilities and agency to develop common pathways using the SDGs as enablers of change", to "produce collective knowledge", and to "progressively transform local institutions and support the evolution of multilevel governance processes" (Bilsky *et al.* 2021, p. 713).

Inclusive, participatory approaches are thus central and "at the heart of smart city development" (Dolmans *et al.* 2023, p. 1577), and in this way, smart cities "have been developing as bottom-up projects, bringing together smart initiatives driven by public bodies, enterprises, citizens, and not-for-profit organizations" (Dameri and Benevolo 2016, p. 693). However, smart cities apply their own governance frameworks, there is a lack of consolidated standards on how to consider the perspectives of multiple actors and smart city dialogues cannot act as a panacea (Aleksandrov *et al.* 2022, p. 142, Dameri and Benevolo 2016). While creative problem-solving and innovation flourish "when different experiences, views and ideas complement and disturb each other" (Dolmans *et al.* 2023, p. 1580), collaboration is particularly encouraged by and through actors with similar interests and values. However, as the literature on multi-actor partnerships has also pointed out, collaboration among multiple and diverse actors is considered very challenging due to fundamental differences in organisational settings and institutional logics. In addition to the frequent "opt[ing] out of collaborations long before the contracts end" (and the respective failure of delegation), "high levels of uncertainty in terms of goals, processes and outcomes" are highlighted, making it difficult for a variety of actors "to coordinate and align on project strategy" (Dolmans *et al.* 2023, p. 1578). But how to overcome such challenges?

As smart city research is still in its infancy, to address the complex organisational challenges associated with the smart city-approach scholars call for more "dialogue between organization studies and smart city research" (Mora *et al.* 2023, p. 1560). In this way and drawing on institutional theory, Dolmans *et al.* (2023 p. 1579) developed a causal loop model to explain why collaborative partners "vary in their adherence to institutional logics over time, which […] greatly influences the collaborative dynamics". They found that it is not only different public and private sector logics that explain the dynamics of smart city development, but also the interplay of institutional logics with uncertainty and governance complexity (Dolmans *et al.* 2023) also supporting the relevance attributed to the internal, external, and temporal dimensions acknowledged in the previous section. Their findings further support orchestration over delegation as the preferred mode of governance for processes aimed at advancing smart city development and the co-creation of innovative solutions, as they involve the recombination of knowledge, which requires a collaborative mode of innovation rather than a bureaucratic or hierarchical one.

To advance policy transfer and alignment (2) in the face of complex problems such as climate change and sustainable development, scholars have focused on multilevel governance as an analytical framework (Gordon 2016, Homsy *et al.* 2019) and have also developed concepts to anticipate opportunities and challenges for policy transfer across governance levels (Hulicka *et al.* 2023). In this way, subnational research (SNR) and MLG share the understanding that "actors and institutions located at one territorial level are shaped by and shape other levels of government" (Giraudy and Niedzwiecki 2022, p. 393). By using the "Policy Transfer Across Governance Systems" heuristic on city-scale climate change policy, for example, the researchers identified "problem identification; the type of transfer networks used; and what aspects of policy contents and goals can be transferred" as three key and dynamic complexities that actors need to address in their respective policy-making across governance levels (Hulicka *et al.* 2023, p. 4). Homsy *et al.* (2019, p. 7) further stressed the need of "[a] sanctioning and coordinating authority, provision of capacity, knowledge co-production, framing of co-benefits, and engagement of civil society" to advance policy alignment. Research on capability approaches that seek to improve institutional capacity to deliver sustainable development also emphasised the need for a central coordinating authority to address resource inequalities and, based on lessons from EU regional policy, a *central coordinating authority* with "the power to enforce decentralized actors' compliance, mobilize the implementation bodies, and provide actors with adequate organizational structures and the resources that they lack". This approach is seen as a prerequisite for effective policy implementation across governance levels (Casula 2022, p. 1). In this way, research on sustainable development has shown how orchestration is already applied by national governments that built regulatory and governance relationships to coordinate and "reorder economic, social and ecological challenges and devolve responsibilities at the sub-national level" (De Laurentis 2020, p. 3) and used intermediary organisations to build capacities and shape policies (Guerreiro and Botetzagias 2018).

To facilitate the collaboration of diverse actors, research on sustainability and urban development underlines that *the institutional set-up and leadership* (3) significantly matter. A governing unit, such as an alliance or multi-actor partnership, needs to acknowledge different institutional logics under which collaborating actors operate (including different rules, norms, knowledge, and discourses) and—from a management perspective—develop respective guidelines "as frames of reference to guide and give meaning to their activities" within this institutional complexity (Dolmans *et al.* 2023, p. 1581). When considering how leadership should ideally look and when different forms of leadership should be in place (see the previous section), scholars more often acknowledge the collective turn in leadership research according to which "individual and collective forms of leadership can coexist, interact and potentially support one another" (Empson *et al.* 2023, p. 202).

The collaborative process of multiple actors in smart city development, however, is further perceived as "complex, non-linear and iterative" in which "micro-level dynamics that characterize collaborative innovation efforts for smart city development" matter as they can turn out to be enablers and barriers to smart city development (Dolmans *et al.* 2023, p. 1594). Moreover, in this process, "the various actors involved in smart city development may behave differently over the various stages of the collaboration", and they may perceive the same problem differently (also over time) due to their background (Dolmans *et al.* 2023, p. 1595), which is why the inclusion of diverse perspectives in collaborative processes may also challenge individual and collective imaginaries. Based on this observation, Dolmans *et al.* (2023, p. 1595, original emphasis) support orchestration as a mode of governance for pursuing smart city development when concluding that "high levels of uncertainty may *promote* a virtually unlimited solution space as actors are less bound to their own institutional logics—thereby generating a huge innovation potential".

In addition to leadership, the roles taken by leaders matter. In this way and based on different identity narratives, scholars identified multiple leadership archetypes[24] "representing their differing senses of themselves as leaders and their alignment with the organizational narrative of collective leadership" (Empson *et al.* 2023, p. 201). In line with the leadership techniques that orchestrators may apply (Abbott *et al.* 2021, as described in the previous section), Dolmans *et al.* (2023) also draw attention to the different roles that public leaders and managers may assume when they engage with multiple actors. Specifically, they describe "conveners", "facilitators", and "catalysts" as roles that enhance "drivers of collaborative innovation" and help to "partially overcome" collaborative barriers, such as tensions and conflicts of interest in multi-actor settings or uncertainty and unknown dynamics (Dolmans et al. 2023). For the managing actors involved, this requires that they fulfil their roles differently than usual. Elected politicians who engage with other actor groups to discuss innovative solutions, for example, "need to engage in risk-taking behaviour, which they tend to avoid because any failure might harm their reputation" (Sørensen and Torfing 2011, Wegrich 2019 cited Dolmans *et al.* 2023, p. 1581) or limit their prospects for securing re-election.

Adding an additional dimension to the four types of leadership in polycentric governance sketched by Liefferink and Wurzel (2018) as introduced in the previous section is that those who are managing collaborative smart city projects "should have substantial discretion and authority to resist major institutional pressures" (Dolmans *et al.* 2023, p. 1596). In this way, pioneering, defined by Liefferink and Wurzel (2018) as carrying out activities that others have deliberately chosen to follow, is also related to the "more 'flexible' (mission-driven) innovation approach" that Dolmans *et al.* (2023, p. 1596) suggest for those initiating collaborative smart city projects as it allows "for alternative scenarios and unexpected outcomes" and thus paves the way for (radical) innovation and gives room for exploratory activities instead of following "short-term driven exploitative goals" (Dolmans *et al.* 2023, p. 1596). These findings further support the positive understanding of power as an enabling force for organisational change (van Baarle *et al.* 2024).

When considering the external dimension, however, the power of those managing (or orchestrating) smart city projects is limited as "some collaborative barriers grow in a self-reinforcing manner [and] can only be addressed at the level of collaborative systems and networks" (Dolmans *et al.* 2023, p. 1582). This understanding supports the *polycentric and network perspectives* (4) introduced in the previous section, as well as the multi-level perspective applied in urban studies investigating the factors that determine the governance strategies that cities adopt. They emphasise that "divergent arrays of jurisdictional capacity (linked to multilevel distributions of state power) influence how city governments engage with other governance actors" (Noring *et al.* 2021, p. 1343). This also relates to the observation that city-networks still appear "limited in their efforts to govern" (Gordon 2016, p. 529), even though "[n]etworked urban governance is emerging as a major feature of metropolitan strategy and activity" (Davidson *et al.* 2019, p. 697), and it is widely acknowledged that "the global governance of climate change must pass through cities" (Gordon 2016, p. 529) and that the subnational level is crucial for pursuing the global goals and targets.

Platforms, moreover, that bring together diverse actors (e.g., city networks with national governments) seem crucial to strengthening a transfer of local perspectives in policy-making processes for pursuing the global goals and targets, to advance policy alignment and to support innovation. While two decades ago scholars were unsure about whether "we might move through 'governance in the shadow of government' to self-organizing policy and service delivery systems—'governance without government'" (Bovaird 2005, p. 217), it seems that hybrid platforms will increasingly turn into "a fourth mode of governance" (Haveri and Anttiroiko 2023, p. 3). This platform logic supports and is supported by a polycentric perspective and "broadens the view of network governance to a broader set of connections, the orchestration of multiple logics and ecosystem thinking" (Haveri and Anttiroiko 2023). In this way, a central question discussed in the literature on transnational dynamism of cities is how cities and city-networks can raise their profiles within the still-state-centric global governance architecture (Martinez 2022). Here, unsurprisingly, the language applied in the global agreements is considered as a supportive frame that connects the local and the global scales and "the localization of the UN global agendas [as] a narrative that organizes the networked orchestration of the political agency of cities in the global urban age" (Martinez 2022).

Similar to research on cross national networking and city networks, the discussion on platforms in view of urban development is still in its infancy (Davidson *et al.* 2019). Nevertheless, "in connection with the smart city discourse", platforms have received increasing attention in academic debates (Haveri and Anttiroiko 2023, p. 4) as issues on knowledge dynamics, participation and (economic and political) power in particular intersect. Also, more generally, "[u]rban platforms have many characteristics of governance networks" (Haveri and Anttiroiko 2023, p. 14),[25] an idea which supports their characterisation as polycentric governance arrangements that are governed through orchestration (as introduced in the previous section). There are differences, however, among networks and platforms that are particularly visible in both of their formation phases: While actors usually engage in networks due to resource dependencies or shared interests, those who are entering into platforms often seek opportunities for collaboration. Later, when collaborating more intensively, "mutual dependencies may grow and occasionally result in the development of shared goals" (Haveri and Anttiroiko 2023, p. 15). In that way, platforms are seen more as "socio-technical spaces that are designed to collect resources, attract actors and create public value by resource orchestration and the facilitation of collaboration and transactions" (Haveri and Anttiroiko 2023, p. 15).

By exploring platforms as a mode of governance, the still-limited research on urban platforms draws attention to the "dynamics of local platform governance with special regard to the roles and relations of city governments, citizens, and local businesses" (Sahamies *et al.* 2022, p. 1710). Based on evidence from five Finnish platforms, city governments are seen as adopting a "facilitative and enabling role on the platforms" that "seek to create value by utilizing skills, knowledge, and resources of local communities in different kinds of co-creation processes" (Sahamies et al. 2022). It has also been investigated how local authorities as platform owners should "convene, coordinate or administer platforms" and how the role of platforms should be conceived in urban governance (Haveri and Anttiroiko 2023, p. 4). These questions relate to the internal and external dimensions in the operating environment of urban platforms, which—as explained in the previous section—differ from platform to platform depending on their setting (territorial varieties) and set-up. The formation of platforms (platformisation) is thus shaped by multiple factors, of which urban density, local embeddedness and paradigm shifts are examples. In view of the latter, for example, the coordination in public administration traditionally "relied almost entirely on hierarchies" (Haveri and Anttiroiko 2023, p. 5). However, the New Public Governance paradigm promoting the "common good" and the inclusion of "public values" across the political system "facilitates an understanding of inter-connections, inter-dependencies, interactions between complex issues and across multiple boundaries, to reach agreement between diverse stakeholders influencing what constitutes 'public value'" (Liddle 2021, p. 1). In this way, networks[26] are considered key in contemporary public service delivery, and, through platforms, these coexisting networks are managed (orchestrated) by the city government to provide public services (Millard 2018 cited Haveri and Anttiroiko 2023).

Overall, the findings presented above show that the local level and participatory approaches are seen as most relevant in the smart city discourse for pursuing the global goals in cities, but that collaboration between multiple actors is seen as challenging due to differences in organisational settings, institutional logics, high levels of uncertainty, and governance complexity (1). To address these challenges, organisational studies have highlighted the need for a central coordinating authority, such as an orchestrator, who also helps facilitate policy transfer and alignment (2). However, this authority needs to recognise differences in institutional arrangements and adapt its leadership over time to the dynamics of collaboration. More flexible forms of governance ideally succeed in overcoming barriers to collaboration and enabling different types of actors to work together (3). From a polycentric and network perspective, however, the governance efforts of city networks and platforms are still seen as limited, even though platformisation is encouraged by the language used in global governance instruments and the New Public Governance paradigm (4). While the smart city discourse in particular has driven the attention given to platforms in academic debates, whether or not platforms will also become a "fourth mode of governance" in urban development, although likely, remains to be seen. In view of the research presented in this book, however, "the platform logic broadens the view of network governance to a broader set of connections, the orchestration of multiple logics and ecosystem thinking" (Haveri and Anttiroiko 2023, p. 17) and supports the idea of direct or participatory democracy (Sahamies *et al.* 2022).

Seeking to better understand and effectively operationalise the theoretical perspectives discussed above, we next introduce the conceptual model representing how local approaches to SUD in remote regions are shaped and how they can be advanced. It includes five critical components responsible for how visions, policies and decisions are framed and introduced in policy- and decision-making processes on SUD at the local level. These components cover three interlinked factors: (1) actors and their relationships, (2) institutions and their set-ups, and (3) political priorities. Additionally, the model includes two key drivers, namely, (4) imaginaries and (5) cooperation, that influence the interlinked factors. The visual representation of the model is presented in Figure 3.1 (Chapter 3). All factors and key drivers are place- and time-sensitive and may have different impacts depending on the setting.

The first factor considers the (vested) interests that arise from the actors involved; their roles and dependence on electoral cycles; and their membership in different alliances, position, profession, and personal perspectives. The second factor derives from the notion that visions of SUD are embedded in local institutional set-ups and shaped by available and accessible capacities, which also affect the political, legislative, and scientific approaches applied to pursue them. The directions, agendas, and priorities in urban development, which are determined by the relational and temporal character of the concept of "sustainable development" itself, are covered by the third factor. In our understanding, the factors are subject to change (they adjust to achieve specific outcomes), which means that they are the results of independent or unilateral political decisions. Thus, the first key driver draws upon the understanding that visions of "sustainable development" are embedded in imaginaries of the geographical environment (Steinberg *et al.* 2015) associated with the adopted system of social values and identity,[27] and the second key driver considers that visions can be stimulated by or are connected with the possibilities for cooperation (Leal Filho *et al.* 2022), which are always dependent on the other actors' decisions. These factors and key drivers interlink with each other as the features ascribed to them might interact in practice. For instance, different formats of cooperation shape the roles ascribed to actors involved in urban development (e.g., recognition of the position of Indigenous peoples on the Arctic Council as Permanent Participants). Likewise, the institutional set-ups and cities' capacities affect their level of cooperation. Moreover, we are aware that these factors and key drivers are not exclusive; however, we consider these five as central to how local approaches to SUD in remote regions are developed. As such, we argue that how these three factors and two key drivers are adjusted at the local level significantly affects the pursuit of the global goals. A detailed presentation and explanation of the conceptual model is provided in the following chapter.

Acknowledgements

This chapter was written by Dorothea Wehrmann. The outline was developed by Michał Łuszczuk and Dorothea Wehrmann. The chapter was reviewed by Jacqueline Götze, Arne Riedel and an external expert. The graphics were designed by Katharina Schaarschmidt (IDOS). All authors of this book read and accepted the content of this chapter.

Notes

1 The Paris Agreement emphasises the "intrinsic relationship that climate change actions, responses and impacts have with equitable access to sustainable development" (UNFCCC 2016, p. 2). The 2030 Agenda similarly acknowledges the need to "minimize the impact of cities on the global climate system" and it recognises "that sustainable urban development and management are crucial to the quality of life of our people". It further declares "to work with local authorities and communities to renew and plan our cities" (UNGA 2015). The New Urban Agenda builds on the aim to localise the 2030 Agenda (UN 2017).

2 Cf., that is, 2030 Agenda (SDG 17.16): "Enhance the global partnership for sustainable development, complemented by multi-stakeholder partnerships that mobilize and share knowledge, expertise, technology and financial resources, to support the achievement of the sustainable development goals in all countries, in particular developing countries" and "We will promote the systematic use of multi-stakeholder partnerships in urban development processes, as appropriate, establishing clear and transparent policies, financial and administrative frameworks and procedures, as well as planning guidelines for multi-stakeholder partnerships" (UN 2017, p. 39).

3 The 2030 Agenda, for instance, seeks "to foster inter-cultural understanding, tolerance, mutual respect and an ethic of global citizenship and shared responsibility" and further "recognize[s] that all cultures and civilizations can contribute to, and are crucial enablers of, sustainable development" (Article 36). In its description of SDG 11, it further stresses enhancing "inclusive and sustainable urbanization and capacity for participatory, integrated and sustainable human settlement planning and management in all countries" (SDG 11.3). Also, the parties to the New Urban Agenda "commit ourselves […] to working with local communities and local governments to identify opportunities for engaging and developing local, durable and dignified solutions" (UN 2017, p. 12). They further promote "institutional, political, legal and financial mechanisms in cities and human settlements to broaden inclusive platforms, in line with national policies, that allow meaningful participation in decision-making, planning and follow-up processes for all, as well as enhanced civil engagement and co-provision and co-production" (UN 2017, p. 14).

4 While the 2030 Agenda derives from a people-centric approach ("It is an Agenda of the people, by the people, and for the people", Article 52) and "tak[es] into account different national realities, capacities and levels of development", it distinguishes only between "developed and developing countries" (Article 5). Likewise, the Paris Agreement considers "the urgent and immediate needs of those developing country Parties that are particularly vulnerable to the adverse effects of climate change" (Article 7, p. 9). An exception is, however, Article 22 in the 2030 Agenda (emphasis added) which differentiates between economic and geographic factors: "Each country faces specific challenges in its pursuit of sustainable development. The most vulnerable countries and, in particular, African countries, least developed countries, landlocked developing countries and small island developing states deserve special attention, as do countries in situations of conflict and post-conflict countries. *There are also serious challenges within many middle-income countries*".

5 In more concrete terms, the New Urban Agenda seeks to "support local government associations as promoters and providers of capacity development, recognizing and strengthening, as appropriate, both their involvement in national consultations on urban policies and development priorities and their cooperation with subnational and local governments, along with civil society, the private sector, professionals, academia and research institutions, and their existing networks, to deliver on capacity-development programmes. This should be done by means of peer-to-peer learning, subject-matter-related partnerships and collaborative actions, such as inter-municipal cooperation, on a global, regional, national, subnational and local scale, including the establishment of practitioners' networks and science-policy interface practices" (UN 2017, p. 37).

6 The 2030 Agenda states, that is, "Our Governments have the primary responsibility for follow-up and review, at the national, regional and global levels, in relation to the progress made in implementing the Goals and targets over the coming fifteen years" (Article 47).

7 Multi-actor partnerships consist of actors from at least three sectors (governmental, non-governmental or the private sector; Partnerships 2030, 2017). Multi-actor partnerships are mostly based on agreements (not on legally-binding contracts). They may address different target groups (Biekart and Fowler 2016), base their cooperation on different mandates and objectives (e.g., setting of standards, knowledge-sharing, provision of services, Beisheim, Janetschek, and Sarre 2014, Pattberg and Widerberg 2014), and differ in their structures, such as the levels (local, regional, global) at which they primarily operate (Loveridge and Wilson 2017, Treichel et al. 2016).

8 This question relates not only to the individual or organisational level of leadership but also considers diverse thematic contexts that may require leadership at different scales. Climate leadership, for example, "requires leadership across both sectors and governance levels" (Ostrom 2009, Torney 2019 cited in Benulic *et al.* 2022). In polycentric systems of governance; moreover, "multiple policy issues, that is, topical policy areas or problems that they address, affect, or both" are drawn together, which require different forms of coordination that may change over time (Morrison *et al.* 2023, p. 478).

9 In contrast, network theory can be considered as "the structural backbone of a complex system" (Kim 2020, p. 919), such as polycentric governance theory, which focuses on the structure of relationships but not on the dynamic processes in polycentric networks, which are "highly clustered at the node level, modular at the community level, and decentralized at the network level" (Kim 2020, p. 917).

10 The abbreviations that we use for the different linkages were not introduced by Pattberg *et al.* but are abbreviations that we use for our research purposes only.

11 In this way, also the Paris Agreement can be seen as "an integrative device" where "normative foundations of the UNFCCC are streamlined into non-state initiatives" (Pattberg *et al.* 2018, p. 184).

12 Different to leaders, pioneers carry out activities that others deliberately chose to follow while leaders carry out activities together with other followers that they lead (Liefferink and Wurzel 2018).

13 All actors may exert entrepreneurial leadership, which "involves diplomatic and/or lobbying efforts, ...[...] initiating a strategic lawsuit or sharing knowledge in a network of peers", and "allows actors of limited size and capacity to exert leadership far beyond the boundaries of their own polycentric unit" (Liefferink and Wurzel 2018, p. 147).

14 One approach that advances the understanding of internal dynamics in polycentric governance is, for example, the "building-blocks approach" (Morrison *et al.* 2023).

15 For concrete examples on how polycentricity can be applied see Thiel *et al.* (2019).

16 As Morrison *et al.* (2023) point out: "The success of polycentric governance is dependent upon how well actors can overcome numerous and interlinked collective action challenges". They further differentiate between horizontal and vertical coordination, with the former referring to the coordination of "resources and information across ecologically and socially diverse landscapes, and cross-level" and the latter to "coordination from local to global levels of policymaking", which both "require organization and cooperation among heterogeneous stakeholder groups". In this way, Chan *et al.* (2021, p. 27) further suggest to combine insights on emerging polycentric structures "with tools that map (goal) coherence" as "[t]he combination of these fields of knowledge could inform supportive policies, for instance in development cooperation to ensure greater coherence in implementing sustainable development priorities".

17 With delegation, control is maximised, while with orchestration it is "wholly non-hierarchical: the intermediary neither owes its authority nor risks losing it to the orchestrator" (Abbott *et al.* 2021, p. 141). Trusteeship and cooperation are other modes of indirect governance but the competence-control trade-off, which "is crucial to understanding principals' choices among indirect governance modes" differs the most between delegation and orchestration (Abbott *et al.* 2021).

18 Subnational governments in Scandinavian countries, for example, enjoy "extensive tax-raising and borrowing autonomy" but have "fairly little political powers", and in Austria it is the reverse (Enderlein 2010, p. 427).

19 "The concept PPP relates to contractual arrangements between the state and a private-sector company. These arrangements define how the private sector participates in the supply of goods, assets and services normally provided by the public sector and how risks are shared (following Ion, Beyard, and Sedaca 2014, Romero 2015, p. 4)" (Wehrmann 2018).

20 Under the auspices of the Arctic Council, the Arctic states negotiated three legally binding agreements: The Agreement on Cooperation on Aeronautical and Maritime Search and Rescue in the Arctic (signed 2011), the Agreement on Cooperation on Marine Oil Pollution Preparedness and Response in the Arctic (signed 2013), and the Agreement on Enhancing International Arctic Scientific Cooperation (signed 2017).

21 Others have argued that "[i]t is not fragmentation per se, but rather the (lack of) coordination of fragmented or differentiated institutions, that is a problem for global governance" (Zürn and Faude 2013, p. 120).

22 Chan *et al.* (2018) differentiate between indirect and direct catalytic linkages. Direct catalytic linkages occur "when orchestrators enlist intermediaries, who in turn target more actors" while indirect catalytic linkages "concern the influence beyond the original programmatic orchestration, for instance, when orchestrators mimic or adjust mobilization efforts in response to other orchestrators; when orchestrators encourage participation in other orchestration efforts than their own" (Chan *et al.* 2018, p. 137–138).

23 "The paradigm of Smart Cities arises as a response to the goal of creating the city of the future, where (1) the well-being and rights of their citizens are guaranteed, (2) industry and (3) urban planning is assessed from an environmental and sustainable viewpoint. Smart Cities still face some challenges in their implementation, but gradually more research projects of Smart Cities are funded and executed. Moreover, cities from all around the globe are implementing Smart City features to improve services or the quality of life of their citizens" (Sánchez-Corcuera *et al.* 2019).

24 Empson *et al.* (2023) identify multiple individual leadership archetypes that are present in "any organization with a strongly professed narrative of collective leadership"

(Empson *et al.* 2023, p. 223): the Avatars, who lead by embodying the collective; the Servants, who lead by sacrificing themselves for colleagues (servant); the Sages, who lead by personifying wisdom; the Intrapreneurs, who lead by initiating internally oriented change; the Entrepreneurs, who lead by initiating externally oriented change; the Performers, who lead by role-modelling achievement and the Challengers, who lead by disrupting conformity.

25 "Platforms cluster around a set of resources, which emerge, are sustained and are changed through a series of interactions. Despite their high degree of autonomy and market-like coordination mechanism, they also orchestrate resources and facilitate transactions—which comes near to network management—in order to deal with the different values, operational logics and preferences that are inherent in urban platforms" (Haveri and Anttiroiko 2023).

26 In these networks, "actors are brought together by resource interdependence, which, together with trust and reciprocity, is the underlying rationale for collective action and the pursuing of common goals. As a mode of governance, networks are characterized by dynamism, a flat form, informality and flexibility" (Haveri and Anttiroiko 2023, p. 5).

27 Moreover, imaginaries are at play at all governance levels and also visible in the global agreements, which is why we see very undifferentiated understandings of "emerging economies", "the Global South", and "remote regions" in global agreements. These undifferentiated understandings are also imaginaries that shape how governments at the national levels approach remote regions and cities in their respective national contexts (e.g., via reporting mechanisms such as the Voluntary National Reviews for the UN High-Level Political Forum). They also shape the national relationship between the South and the North.

References

Abbott, K. W., Genschel, P., Snidal, D., and Zangl, B., 2012. Orchestration: Global governance through intermediaries. Available from: https://papers.ssrn.com/sol3/papers.cfm?abstract_id=2125452 [Accessed 14 October 2021].

Abbott, K., Genschel, P., Snidal, D., and Zangl, B., 2016. Two logics of indirect governance: Delegation and orchestration. *British Journal of Political Science*, 46 (4), 719–729. DOI:10.1017/S0007123414000593.

Abbott, K. W., Genschel, P., Snidal, D., and Zangl, B., 2018. The governor's dilemma: Competence versus control in indirect governance. WZB Discussion Paper No. SP IV 2018–101.

Abbott, K. W., Genschel, P., Snidal, D., and Zangl, B., 2020. Competence versus control: The governor's dilemma. *Regulation & Governance*, 14 (4), 619–636. DOI:10.1111/rego.12234.

Abbott, K. W., Genschel, P., Snidal, D., and Zangl, B., 2021. Orchestration: Global governance through intermediaries. *In*: Abbot, K. W. and Snidal, D. J., eds., *The spectrum of international institutions*. London: Routledge, 140–170.

Adams, B., and Martens, J., 2015. Fit for whose purpose? Private funding and corporate influence in the United Nations. Bonn/New York. Available from: https://sustainabledevelopment.un.org/content/documents/2101Fit_for_whose_purpose_online.pdf [Accessed 8 September 2018].

Agenda 2030 2015. Transforming our world: The 2030 agenda for sustainable development. UN General Assembly. Available from: https://www.refworld.org/docid/57b6e3e44.html [Accessed 7 March 2023].

Aleksandrov, E., Dybtsyna, E., Nazarova, N., and Khodachek, I., 2022. Smart city dialogue in the Arctic: Opportunities and challenges. *In*: Mineev, A., Bourmistrov, A., and

Mellemvik, F., eds., *Global development in the Arctic: International cooperation for the future*. London: Routledge, 131–146. DOI:10.4324/9781003246015-10.

Andonova, L. B., Faul, M. V., and Piselli, D., 2022. Partnerships for sustainability in contemporary global governance: Pathways to effectiveness. London: Taylor & Francis.

Bäckstrand, K., Campe, S., Chan, S., Mert, A., & Schäferhoff, M., 2012. Transnational public-private partnerships. In Biermann, F. & Pattberg, P., eds., *Global environmental governance reconsidered*. Cambridge, MA: The MIT Press, 123–148.

Barber, W. F., and Bartlett, R. V., 2021. Democratic governance in the Anthropocene: Equivocal, experimental, equitable, empowered, embedded. *In*: Baber, W. F., and Bartlett, R. V., eds. *Democratic norms of earth system governance. Elements in earth system governance*. Cambridge: Cambridge University Press, 1–22.

Beisheim, M., Janetschek, H., and Sarre, J., 2014. What's the best fit? Partnership project design and its influence on effectiveness. *In*: Marianne Beisheim, A. L., ed., *Transnational partnerships. Effectively providing for sustainable development?* New York: Palgrave Macmillan, 161–192.

Beisheim, M., and Simon, N., 2017. Die Umsetzung der SDGs durch Multi-Stakeholder Partnerschaften. Bessere Meta-Governance seitens der Vereinten Nationen gewünscht? *In*: Lepenies, P., and Sondermann, E., eds., *Globale politische Ziele. Bestandsaufnahme und Ausblick des Post-2015 Prozesses*. Baden-Baden: Nomos, 195–217.

Benulic, K.-S., Kropf, M., Linnér, B.-O., and Wilbeck, V., 2022. The meaning of leadership in polycentric climate action. *Environmental Politics*, 31 (6), 1016–1036, DOI:10.1080/09644016.2021.1970087.

Biekart, K., and Fowler, A., 2016. Effective multi-stakeholder initiatives through civic engagement. *Paper presented at the 12th International Conference of the International Society for Third Sector Research*, Stockholm. Avaialbe from: https://www.researchgate.net/profile/Alan_Fowler/publication/305719346_Effective_Multi-Stakeholder_Initiatives_through_Civic_Engagement/links/579b713b08ae802facba56b8/Effective-Multi-Stakeholder-Initiatives-through-Civic-Engagement.pdf [Accessed 2 October 2018].

Bilsky, E., Moreno, A. C., and Fernández Tortosa, A., 2021. Local governments and SDG localisation: Reshaping multilevel governance from the bottom up. *Journal of Human Development and Capabilities*, 22 (4), 713–724. DOI:10.1080/19452829.2021.1986690.

Bovaird, T., 2005. Public governance: Balancing stakeholder power in a network society. *International Review of Administrative Sciences*, 71 (2), 217–228. DOI:10.1177/0020852305053881

Britton, J., 2019. Smart meter data and equitable energy transitions—Can cities play a role? *Local Environment*, 24 (7), 595–609. DOI:10.1080/13549839.2017.1383372.

Casula, M., 2022. How different multilevel and multi-actor arrangements impact policy implementation: Evidence from EU regional policy. *Territory, Politics, Governance*, 12 (7), 1048–1072. DOI:10.1080/21622671.2022.2061590.

Chan, D. K., 2016. City diplomacy and "glocal" governance: Revitalizing cosmopolitan democracy. *Innovation: The European Journal of Social Science Research*, 29 (2), 134–160.

Chan, S., Ellinger, P., and Widerberg, O., 2018. Exploring national and regional orchestration of non-state action for a < 1.5°C world. *International Environmental Agreements*, 18, 135–152. DOI:10.1007/s10784-018-9384-2.

Chaturvedi, S., *et al.*, 2021. The Palgrave handbook of development cooperation for achieving the 2030 Agenda. Cham: Springer International Publishing.

Coen, D., and Thatcher, M., 2008. Network governance and multi-level delegation: European networks of regulatory agencies. *Journal of Public Policy*, 28 (1), 49–71. DOI:10.1017/S0143814X08000779.

Cruz, S. F., 2023. SDG 17 and global partnership for sustainable development: Unraveling the rhetoric of collaboration. *Frontiers in Environmental Science*. 11. Available from: DOI:10.3389/fenvs.2023.1155828 [Accessed 19 February 2024].

Dameri, R. P., and Benevolo, C., 2016. Governing smart cities: An empirical analysis. *Social Science Computer Review*, 34 (6), 693–707. DOI:10.1177/0894439315611093

Davidson, K., Coenen, L., Acuto, M., and Gleeson, B., 2019. Reconfiguring urban governance in an age of rising city networks: A research agenda. *Urban Studies*, 56 (16), 3540–3555. DOI:10.1177/0042098018816010

De Laurentis, C., 2020. Mediating the form and direction of regional sustainable development: The role of the state in renewable energy deployment in selected regions. *European Urban and Regional Studies*, 27 (3), 303–317. DOI:10.1177/0969776420904989

Dolmans, S. A. M., *et al.*, 2023. A dynamic perspective on collaborative innovation for smart city development: The role of uncertainty, governance, and institutional logics. *Organization Studies*, 44 (10), 1577–1601. DOI:10.1177/01708406231169422

Eichenberger, R. and Frey, B. S., 2006. Functional, Overlapping and Competing Jurisdictions (FOCJ): A Complement and Alternative to Today's Federalism. *In:* Ehtisham, A. and Brosio, G., eds. *Handbook of Fiscal Federalism*. Cheltenham: Edward Elgar, 154–181.

Empson, L., Langley, A., and Sergi, V., 2023. When everyone and no one is a leader: Constructing individual leadership identities while sustaining an organizational narrative of collective leadership. *Organization Studies*, 44 (2), 201–227. DOI:10.1177/01708406221135225

Enderlein, H., 2010. 28 Economic policy-making and multi-level governance. In: Enderlein, H., Wälti, S., and Zürn, M., eds., *Handbook on multi-level governance*. Cheltenham: Edward Elgar, 423–440.

Epstein, D., and O'Halloran, S., 1999. Delegating powers. Cambridge: Cambridge University Press.

Freistein, K., Mahlert, B., Quack, S., and Unrau, C., eds., 2022. Imagining pathways for global cooperation. Cheltenham and Northampton, MA: Edward Elgar.

Fukuyama, F., 2013. What is governance? *Governance*, 26 (3), 347–368. DOI:10.1111/gove.12035.

Giraudy, A., and Niedzwiecki, S., 2022. Multi-level governance and subnational research: Similarities, differences, and knowledge accumulation in the study of territorial politics. *Regional & Federal Studies*, 32 (3), 393–411. DOI:10.1080/13597566.2021.1941900.

Gordon, D. J., 2016. Lament for a network? Cities and networked climate governance in Canada. *Environment and Planning C: Government and Policy*, 34 (3), 529–545. DOI:10.1177/0263774X15614675.

Guerreiro, S., and Botetzagias, I., 2018. Empowering communities—The role of intermediary organisations in community renewable energy projects in Indonesia. *Local Environment*, 23 (2), 158–177. DOI:10.1080/13549839.2017.1394830.

Haveri, A., and Anttiroiko, A.-V., 2023. Urban platforms as a mode of governance. *International Review of Administrative Sciences*, 89 (1), 3–20. DOI:10.1177/00208523211005855

Hawkins, D. G., Lake, D. A., Nielson, D. L., and Tierney, M. J., 2006. Delegation and agency in international organizations. Cambridge: Cambridge University Press.

Heinkelmann-Wild, T., and Mehrl, M., 2022. Indirect governance at war: Delegation and orchestration in rebel support. *Journal of Conflict Resolution*, 66 (1), 115–143. DOI:10.1177/00220027211027311

Homsy, G. C., Liu, Z., and Warner, M. E., 2019. Multilevel governance: Framing the integration of top-down and bottom-up policymaking. *International Journal of Public Administration*, 42 (7), 572–582. DOI:10.1080/01900692.2018.1491597.

Hooghe, L., and Marks, G., 2010. Types of multi-level governance. *In:* Enderlein, H., Wälti, S., and Zürn, M., eds., *Handbook on multi-level governance*. Cheltenham: Edward Elgar, 17–31.

Horner, R., and Hulme, D., 2017. Converging divergence? Unpacking the new geography of 21st century global development (GDI working papers 2017–010). Manchester: University of Manchester.

Hulicka, A., Lucas, P., and Carson, L., 2023. Policy transfer across governance systems: An adapted approach. *Policy Studies*, 44 (1), 4–25. DOI:10.1080/01442872.2021.1987407.

Jakobeit, C., Kappel, R., and Mückenberger, U., 2010. Civilizing the world order? The scope and potential of transnational norm-building networks. Available from: *https://www.files.ethz.ch/isn/114076/gf_international_2010-01.pdf* [Accessed 31 July 2022].

Klingebiel, S., Mahn, T. C., and Negre, M., 2016. Fragementation: A key concept for development cooperation. *In*: Klingebiel, S., Mahn, T. C., and Negre, M., eds., The fragmentation of aid: Concepts, measurements and implications for development cooperation. Basingstoke: Palgrave Macmillan, 1–18.

Kranholdt, D., 2022. Have we failed to capitalise on the momentum for the SDGs? The Current Column. Bonn: German Institute of Development and Sustainability (IDOS). Available from: https://www.idos-research.de/en/the-current-column/article/have-we-failed-to-capitalise-on-the-momentum-for-the-sdgs/ [Accessed 2 February 2023].

Leal Filho, W., Wall, T., and Barbir, J. *et al.*, 2022. Relevance of international partnerships in the implementation of the UN sustainable development goals. *Nature Communications*, 13, 613. DOI:10.1038/s41467-022-28230-x

Léautier, F., 2014. Leadership and governance. *In:* Léautier, F., ed., Leadership in a globalized world. London: Palgrave Macmillan, 126–176. DOI:10.1057/9781137431219_5

Liddle, J., 2021. New public governance. *In*: List, R. A., Anheier, H. K., and Toepler, S., eds., *International encyclopedia of civil society*. Cham: Springer, 1–6. DOI:10.1007/978-3-319-99675-2_9580-1

Liefferink, D., and Wurzel, R., 2018. Leadership and pioneership: Exploring their role in polycentric governance. *In*: Jordan, A., Huitema, D., Van Asselt, H., and Forster, J., eds., *Governing climate change: Polycentricity in action?* Cambridge: Cambridge University Press, 135–151. DOI:10.1017/9781108284646.009

Loveridge, D., and Wilson, N., 2017. Engaging with the private sector through multi-stakeholder platforms. Available from: https://www.enterprise-development.org/wp-content/uploads/DCED-Platforms-Review.pdf [Accessed 5 February 2018].

Martinez, R., 2022. Framing the localization of the global agendas: Orchestrating the political agency of cities within the local-global Nexus. *Alternatives*, 47 (2), 100–114. DOI:10.1177/03043754221095303

McAllister, R. J., and Taylor, B. M., 2015. Partnerships for sustainability governance: A synthesis of key themes. *Current Opinion in Environmental Sustainability*, 12, 86–90. DOI:10.1016/j.cosust.2015.01.001.

Mora, L., *et al.,* 2023. Organizing for smart city development: Research at the crossroads. Introduction to the Special Issue. *Organization Studies*, 44 (10), 1559–1575. DOI:10.1177/01708406231197815

Morrison, T. H., *et al.,* 2017. Mitigation and adaptation in polycentric systems: Sources of power in the pursuit of collective goods. *WIREs Climate Change*, 8 (September/October), 1–16. Available from: https://onlinelibrary.wiley.com/doi/full/10.1002/wcc.479 [Accessed 12 September 2023].

Morrison, T. H., *et al.,* 2023. Building blocks of polycentric governance. *Policy Studies Journal*, 51, 475–499. DOI:10.1111/psj.12492

Nesbit, M., Filipova, T., Stainforth, T., Nyman, J. et al., 2019. *Development of an assessment framework on environmental governance in the EU Member States – Final report*, Publications Office of the European Union. Available from: https://data.europa.eu/doi/10.2779/299476 [24 November 2024]. New Urban Agenda. H III: Habitat III: Quito 17–20 October 2016 (2017). Nairobi: United Nations.

Noring, L., Ohler, L. P., and Struthers, D., 2021. City government capacity and patterns in urban development project governance. *Urban Affairs Review*, 57 (5), 1343–1371. DOI:10.1177/1078087419897808

Nyseth, T., 2017. Arctic Urbanization: Modernity Without Cities. In: Körber, LA., MacKenzie, S., Westerståhl Stenport, A., eds. *Arctic Environmental Modernities. Palgrave Studies in World Environmental History*. Palgrave Macmillan, Cham, 59–70. https://doi.org/10.1007/978-3-319-39116-8_4

Ostrom, E., 2009. Beyond *markets* and states: Polycentric governance of complex economic systems. Paper presented at the Workshop in Political Theory and Policy Analysis, Indiana University, Bloomington, IN, 8 December.

Partnerships 2030 2017. What is an MSP? Available from: https://www.partnerschaften2030.de/en/was-ist-eine-map/#1498753111835-1a44af1b-a883 [Accessed 16 October 2018].

Pattberg, P. H., Chan, S., Sanderink, L., and Widerberg, O. E., 2018. Linkages: Understanding their role in polycentric governance. *In:* Jordan, A., Huitema, D., van Asselt, H., and Forster, J., eds., Governing climate change: Polycentricity in action? Cambridge: Cambridge University Press, 169–187. DOI:10.1017/9781108284646.011

Pattberg, P., and Widerberg, O., 2014. Transnational multi-stakeholder partnerships for sustainable development. Building blocks for success. Available from: *https://papers.ssrn.com/sol3/papers.cfm?abstract_id=2480302* [Accessed 10 October 2018].

Pollack, Mark A., 1997. Delegation, agency, and Agenda setting in the European community. *International Organization*, 51 (1), 99–143.

Prescott, D., and Stibbe, D., 2017. Better together. Unleashing the power of the private sector to tackle non-communicable diseases. A guidebook for collaboration between non-profit organisations and businesses. Available from: https://www.thepartneringinitiative.org/wp-content/uploads/2017/02/Better-Together-Guidebook.pdf [Accessed 10 October 2018].

Rakhyun, E. Kim, 2020. Is global governance gragmented, polycentric, or complex? The state of the art of the network approach. *International Studies Review*, 22 (4), 903–931. DOI:10.1093/isr/viz052

Sahamies, K., Haveri, A., and Anttiroiko, A.-V., 2022. Local governance platforms: Roles and relations of city governments, citizens, and businesses. *Administration & Society*, 54 (9), 1710–1735. DOI:10.1177/00953997211072531

Sánchez-Corcuera, R., *et al.*, 2019. Smart cities survey: Technologies, application domains and challenges for the cities of the future. *International Journal of Distributed Sensor Networks*, 15 (6). DOI:10.1177/1550147719853984.

Steinberg, P. E., *et al.*, 2015 Contesting the Arctic: Politics and imaginaries in the Circumpolar North. London: Bloomsbury Publishing.

Tennyson, R. (2011). The partnering toolbook. An essential guide to cross-sector partnering. Availablefrom:https://thepartneringinitiative.org/wp-content/uploads/2014/08/Partnering-Toolbook-en-20113.pdf [Accessed 13 November 2023].

Thiel, A., Blomquist, W. A., and Garrick, D. E., 2019. Governing complexity. Analyzing and applying polycentricity. Cambridge: Cambridge University Press.

Thorpe, J., and Maestre, M., 2015. Brokering development: Enabling factors for public-private-producer partnerships in agricultural value chains. Available from: https://www.

ifad.org/documents/10180/3533bdf5-ee8c-49a9-9aa5-c3254a3d1507 [Accessed 10 October 2018].

Treichel, K., Höh, A., Biermann, S., and Conze, P., 2016. Multi-Akteurs-Partnerschaften im Rahmen der Agenda 2030. Eine praxisorientierte Analyse von Potentialen, Herausforderungen und Erfolgsfaktoren. Bonn und Eschborn. Available from: https://partnerschaften2030.de/app/uploads/2023/09/Multi-Akteurs-Partnerschaften-im-Rahmen-der-Agenda-2030.pdf [Accessed 2 October 2018].

United Nations 2017. New urban agenda. Available from: https://habitat3.org/wp-content/uploads/NUA-English.pdf [Accessed 10 August 2023].

United Nations 2023. The global sustainable development report (GSDR) 2023. Department of economic and social Affairs. Available from: https://sdgs.un.org/gsdr/gsdr2023 [Accessed 21 November 2023].

United Nations Framework Convention on Climate Change 2016. The Paris Agreement. Available from: https://unfccc.int/sites/default/files/resource/parisagreement_publication.pdf [Accessed 10 August 2023].

van Baarle, S., Bobelyn, A. S., Dolmans, S. A., and Romme, A. G. L., 2024. Power as an enabling force: An integrative review. *Human Relations*, 77 (2), 143–171. DOI:10.1177/00187267221128561

Wehrmann, D., 2018. Incentivising and regulating multi-actor partnerships and private-sector engagement in development cooperation. Discussion Paper 18/2018. Bonn: German Development Institute/Deutsches Institut für Entwicklungspolitik (DIE). Available from: https://www.idos-research.de/en/discussion-paper/article/incentivising-and-regulating-multi-actor-partnerships-and-private-sector-engagement-in-development-cooperation/ [Accessed 15 January 2021].

Wehrmann, D., 2020. The Arctic Council as a success case for transnational cooperation in times of rapid global changes? Arctic Yearbook 2020. Available from: https://arcticyearbook.com/arctic-yearbook/2020/2020-scholarly-papers/358-the-arctic-council-as-a-success-case-for-transnational-cooperation-in-times-of-rapid-global-changes[Accessed 15 January 2021].

Young, O. R., 2021. *Grand challenges of planetary governance: Global order in turbulent times*. Cheltenham: Edward Elgar.

Young, O. R., 2023. Addressing the grand challenges of planetary governance. Cambridge: Cambridge University Press.

Zürn, M., and Faude, B., 2013. Commentary: On fragmentation, differentiation, and coordination. *Global Environmental Politics*, 13 (3), 119–130. DOI: DOI:10.1162/GLEP_a_00186.

Zürn, M., Wälti, S., and Eberlein, H., 2010. Introduction. *In*: Enderlein, H., Wälti, S., and Zürn, M., eds., *Handbook on multi-level governance*. Cheltenham: Edward Elgar, 1–13.

3 Conceptual Model for Understanding Local Approaches to Sustainable Urban Development in a Remote Region

3.1 Introduction to the Conceptual Model

In the preceding section, we scrutinised the 2030 Agenda, the Paris Agreement, and the New Urban Agenda, which are global governance instruments addressing global challenges. While each has a unique objective, their goals and implementation strategies intersect in many areas—they all emphasise global cooperation and the collective action of various actors, including governments, the private sector, and civil society.[1]

We acknowledge that effective implementation of the instruments demands engagement across multiple sectors and at international, national, and local levels, affirming the significant role of multi-level governance (MLG) highlighted by recent research (Westman *et al.* 2019, Russel and Kirsop-Taylor 2022). Our analysis of the three agendas and the related literature reveals that addressing the complex challenges in the agreements requires collaboration across multiple governance levels. There is widespread consensus on the necessity of MLG for an effective, inclusive, and comprehensive strategy (Allain-Dupré 2020). Additionally, this approach allows for the integration of global goals with localised strategies through knowledge transfer (Gonzales-Iwanciw *et al.* 2020).

Based on these findings and recognising the importance of MLG as an approach and orchestration as a process for advancing the implementation of the global agendas, we focus here on the conceptual model for understanding local approaches to sustainable urban development in a remote region. As already explained, we argue that this model helps to understand better (1) the engagement of local actors in the orchestration processes for advancing sustainable urban development in remote regions, (2) their policy-making on sustainable development, and (3) the foundational ideas, perspectives, and principles that shape the understandings and approaches to sustainable development[1] (Bardal *et al.* 2021).

It also provides the analytical potential for further investigation as a lens through which data will be organised and interpreted in the chapters that follow. The conceptual model enables us to:

1 systematically aggregate and categorise data from fieldwork and other sources, such as the Arctic Urban Database,[2]

DOI: 10.4324/9781003284864-3

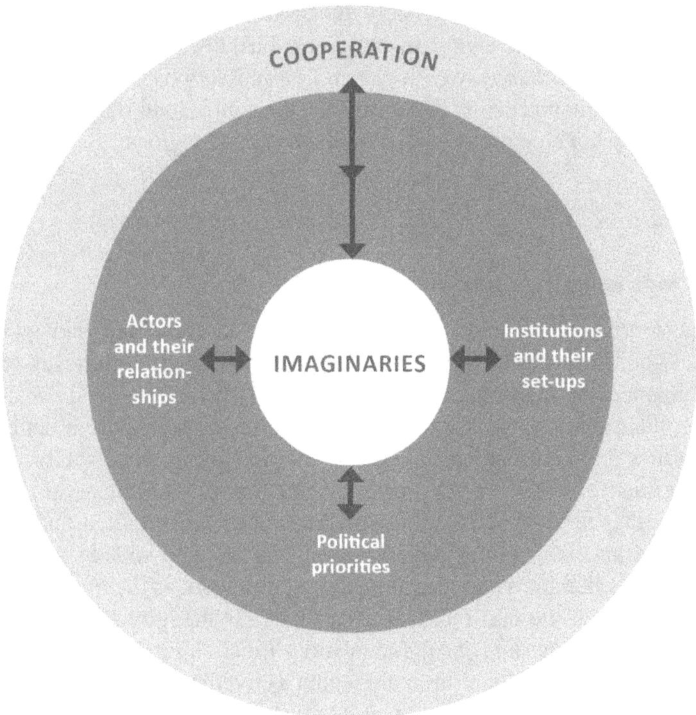

Figure 3.1 Factors and key drivers shaping local approaches to sustainable urban development in remote regions

2 map variations in policy and decision-making processes related to sustainable urban development,
3 to identify gaps in these processes, and
4 deepen our understanding of the challenges in policy alignment and implementation of global agendas at the local level through orchestration.

Applying this model helps us explore why policies on sustainable urban development lack alignment across governance levels, addressing a significant research gap in studies on global agendas and the limitations of MLG (Valencia *et al.* 2019). Moreover, our findings inform the design of pathways (outlined in Chapter 7) for more inclusive local actor participation in policy- and decision-making and enhanced cooperation, which are crucial for achieving global goals.

The conceptual model (see Figure 3.1) enhances current knowledge by integrating insights from a thorough literature review and observations from this research. Our goal is to use this model not only for scientific assessments of sustainable urban development in remote regions but also as a tool that indirectly provides recommendations or guidelines for refining existing strategies or creating innovative, coherent and context-appropriate solutions (see Chapter 9). Furthermore, we emphasise that there is no universally "correct" framework of action; each city must define and follow its own path, balancing local needs with global aspirations.

In the following sections, we discuss in five subsections numerous critical elements of each factor and driver that determine the content and shape of policies and solutions for sustainable urban development in remote regions.

3.2 Factors

3.2.1 Institutional Set-Ups and Capacities

Sustainable urban development is invariably situated and pursued within specific socio-political contexts (see Section 2.1.2) through multilevel structures and is influenced by many factors (James *et al.* 2015, Bulkeley and Betsill 2005). However, the direction and dynamic of SUD policy-making—from comprehending the issues and formulating policies to coordinating various actors, including the engagement of diverse actors, to implementing, monitoring, evaluating, and, finally, adapting new objectives and solutions—can be predicted and controlled (only) to a certain extent (Tang and Lee 2016). It is made possible by a set of issues, suggested in the literature and identified during fieldwork as the main complementary components of the first factor. They include (1) intergovernmental relations, institutional autonomy, and legal frameworks functioning in a state; (2) policy capacities; (3) the institutional structure within a city; (4) citizens' participation and political representation on a local level; (5) financial, infrastructural, and technological resources; and, finally, (6) participation in city networks and partnerships. We explain all of these in detail below.

Intergovernmental Relations, Institutional Autonomy, and Legal Frameworks

Sustainable urban development is a multifaceted undertaking that falls within the scope of competence of specific territorial entities, institutions, or administrative bodies (James *et al.* 2015, Camagni 2017). In this regard, normative and functional measures pertaining to intergovernmental relations and autonomy constitute, alongside sets of legal regulations, pivotal frameworks that significantly affect policy-making, mechanisms for cooperation and negotiation, as well as the balance of power, and ultimately impact the potential for fostering sustainable urban development (Rasoolimanesh *et al.* 2019, Bulkeley and Betsill 2005). A balance between local decision-making autonomy and constructive inter-governmental relations creates a conducive environment for sustainable initiatives, with the potential for far-reaching, long-lasting impacts on urban environments, the local community, and beyond (Bulkeley and Newell 2010).

Inter-governmental relations play a crucial role in coordinating different government levels (federal/national, provincial/territorial/regional/country, local/municipal, and Indigenous). Effective relationships promote policy alignment, reduce overlaps and enhance integrated planning, essential for pursuing sustainability goals. They also facilitate the sharing of financial resources, information and technical expertise vital for sustainable urban development. Supportive legislation from higher government levels can bolster local sustainability efforts. Conversely, strained or conflictual relations can hinder cooperation and lead to gaps in policy outcomes, potentially due to political or systemic issues (Das-Doyle *et al.* 2023).

From an institutional perspective, the autonomy of local institutions significantly impacts sustainable urban development. Institutional autonomy, defined as the independence and self-governance of municipal bodies, such as city councils and urban planning agencies, is essential for streamlined decision-making, efficient resource allocation, and effective accountability. Such autonomy, under democratic pressures, encourages a focus on sustainable and beneficial urban development (Ladner *et al.* 2019, Keuffer and Mabillard 2020). Autonomous institutions are also better positioned to innovate and adapt, trying new strategies for sustainable development that might be constrained in centralised systems. Furthermore, they can integrate urban development policies more effectively with other critical areas such as environment, health, education, and economic development (Medeiros and van der Zwet 2020). In turn, domestic and international legal frameworks critically guide, regulate, and facilitate sustainable urban development (Salet and Vries 2019).

These frameworks provide the standards and tools needed for sustainable growth, though their application often allows for interpretation and can be socially contested. For instance, domestic laws dictate urban planning through zoning regulations, building codes, and environmental impact assessments—all pivotal to sustainability. Environmental laws drive sustainable practices by governing air and water quality, waste management, and conservation efforts. Additionally, property laws affect urban growth and the preservation of green spaces (Camagni 2017). On the international level, agreements such as the Paris Agreement and frameworks such as the New Urban Agenda establish global sustainability goals and standards. While not all international instruments are legally binding, they influence domestic policies and include provisions for pollution control and biodiversity conservation under international environmental law (Mersal 2016). International human rights law also plays a role, protecting Indigenous Peoples' rights and ensuring the right to a healthy environment, thus supporting sustainable urban development efforts. Both domestic and international legal structures are pivotal in shaping urban planning, environmental management, and governance. However, their effectiveness depends on coherence and alignment with local conditions to prevent special interests from hindering sustainable development (Cheshmehzangi and Dawodu 2019).

Policy Capacity

Sustainable urban development is a continuous, iterative process that hinges on the capabilities of all stakeholders involved in designing, implementing, and evaluating effective policies. This involves "policy capacity", which encompasses knowledge, analytical skills, management capabilities, and stakeholder engagement abilities (Pfeffer *et al.* 2013). Policy capacity is crucial at all stages of the policy cycle, including issue understanding, policy formulation, implementation, coordination, stakeholder engagement, monitoring, evaluation, and adaptation. Effective policy formulation for sustainable urban development, for example, demands strong analytical skills to assess data and impacts and to create comprehensive strategies (Sofyani *et al.* 2020). As Bramwell (2020) suggests, high policy capacity also supports innovation in policy design, enabling more practical solutions.

High policy capacity poses unique challenges for cities in remote regions struggling with their demographic characteristics, out-migration processes, limited educational resources, and other issues determined by these areas' particular needs and constraints (Kolehmainen *et al.* 2016). Strengthening local policy capacity for sustainable urban development in a remote context can be achieved via broader community participation and engagement in the local debates; adaptive policy-making; regular assessments and updates; training of members of the local governments and community on sustainable practices, project management, data analysis, etc.; pooling available resources, expertise, and knowledge; stimulating knowledge sharing and building up higher education facilities (Baud *et al.* 2014, Sodiq *et al.* 2019).

Institutional Structure

The institutional structure of city authorities significantly influences the policy-making capacities for sustainable urban development in remote regions. This structure defines the roles and responsibilities of various city government departments, including oversight mechanisms and resource allocation. Clear responsibilities enhance collaboration, coordination, and the implementation of sustainable practices across departments, helping to avoid policy overlaps or gaps (Baud *et al.* 2021).

Moreover, an institutional structure that supports participatory decision-making can integrate diverse local perspectives, leading to more equitable and well-conceived decisions. A transparent and accountable city administration, open to external stakeholders, is crucial for maintaining public trust and effectively implementing sustainability policies. Additionally, adaptable institutional structures enable cities to respond effectively to contemporary challenges, technological changes, and climate impacts, thereby supporting urban resilience and sustainability (Kagan *et al.* 2018). For remote cities, often constrained by limited resources, institutional structures may need to focus narrowly on critical local issues.

Citizens' Participation and Political Representation

Citizens' participation and political representation (mainly by political parties or local independent electoral association) play significant roles in shaping sustainable urban development (Lissandrello *et al.* 2023). Before discussing these issues in more detail in the next section, it is worth pointing out that community engagement in urban planning and development often results in more effective and sustainable solutions. Citizens typically have a strong understanding of their community's needs and challenges; they also often possess local knowledge(s) and have innovative ideas (Berman 2017).

Likewise, citizen participation enriches decision-making by incorporating a wider range of perspectives, leading to more balanced and inclusive outcomes (Åström 2019). Engaging citizens in the policy process increases social acceptance and successful implementation, as policies are viewed as legitimate and reflective of a community's needs. Involvement also heightens accountability in urban development, allowing residents to monitor and evaluate progress, enhancing transparency, and governance legitimacy (Bouzguenda *et al.* 2019). Additionally, including marginalised groups, such as economic immigrants or refugees, ensures a more equitable distribution of urban development benefits, which is vital for sustainable development (Ruiz-Mallén *et al.* 2022).

Citizen engagement in sustainable urban development often takes on political dimensions, highlighting the influential role of political parties. Locally, political parties shape agendas and provide the political will essential for sustainable urban development, playing a key role in formulating and influencing local or regional policies and resource allocation (Bisogno *et al.* 2023). Nationally and internationally, such as within the EU, political parties raise awareness about sustainable urban development. They are pivotal in crafting legislation, influencing budget allocations, and guiding public policies towards sustainability goals (Bossuyt and Savini 2018).

Financial, Infrastructural, and Technological Resources

Sustainable urban development is a sphere of activity that cannot exist without adequate resources, including human capital, social resources, financial resources, access to or offering sufficient services and infrastructure, and digital connectivity for all. Sustainable urban development equally hinges on the strategic management of human capital and the optimisation of social resources. Human capital, which encompasses the education, skills, and health of the urban population, is fundamental to driving innovation, productivity, and resilience in the face of environmental and economic challenges, particularly in remote regions (Petrov 2017). Cities that invest in their inhabitants' education and health are better equipped to develop and implement sustainable practices, as a well-educated workforce is crucial for the research, development, and effective adoption of green technologies and sustainable urban planning strategies (Yiwei and He 2020).

Social resources, including community networks; partnerships among public, private, and non-profit sectors; and civic engagement, play a pivotal role in shaping sustainable urban development. These resources foster collaboration and knowledge sharing, which are essential for addressing complex urban challenges. Engaged communities are more likely to support and participate in sustainability initiatives, such as recycling programmes, community gardens, and local sustainability education efforts (Ehnert *et al.* 2022). Moreover, strong social networks and partnerships can facilitate the mobilisation of resources, both financial and non-financial, towards sustainability projects, enhancing the city's capacity to implement and sustain these initiatives over time (Manzi *et al.* 2015).

Financial resources cover (1) capital investments to support the development of eco-friendly infrastructure, green spaces, efficient public transportation, and renewable energy systems; (2) operational costs needed to maintain and operate urban services sustainably; (3) funding to stimulate the research, development, and adoption of new technologies and practices; and (4) incentives for sustainable behaviour, such as subsidies for renewable energy or grants for energy-efficient building upgrades (Xue *et al.* 2020). In general, for cities in remote regions, it is more difficult to access all these kinds of financial resources than for cities located more centrally, due to smaller populations, lower economic activity, higher costs of doing business, inadequate infrastructure, limited access to financial institutions or development programmes, and reduced level of representation or advocacy, impacting their ability to secure public funds via political influence.

In the same context, that is, geographical location and large distances from other centres, the issues of resources understood as access to the city itself (time/distance/costs of travel) or the possibility of offering citizens adequate and affordable services and infrastructure should be considered as impacting policy-making on sustainable urban development (Pandit *et al.* 2017). These include:

- efficient, accessible, and affordable public transportation (reducing dependence on private vehicles, thereby reducing carbon emissions and air pollution);
- access to public open spaces or green spaces to enhance the quality of life, provide habitats for wildlife, and mitigate the effects of climate change in the most suitable ways depending on the local situation and requirements; and
- sustainable water and waste management services essential for preserving natural resources and reducing pollution.

Finally, the third resource group includes new technology and communication tools supporting digital connectivity, which is particularly vital for remote locations. In this context, it is straightforward to mention the idea of a Smart City based on technologies such as the Internet of Things (IoT), artificial intelligence (AI), and Big Data enabling smarter and more efficient urban services, starting from smart grids for energy to intelligent transportation systems (Ahad *et al.* 2023). Moreover, digital connectivity enables better communication between governments and citizens, fostering greater participation in urban (and not only this) decision-making and increasing transparency and accountability (Volpi *et al.* 2015). Recent years have also proved that digital connectivity facilitates telecommuting and provides online services (e-services), which can reduce transportation needs and associated emissions.

However, the advancement of technology and communication tools brings significant challenges, notably misinformation in the public sphere and the digital divide. These issues have complex implications for society, governance, well-being, and sustainable urban development. The digital divide—the gap in access to and effective use of information and communication technologies—disproportionately affects remote areas due to the high costs and logistical challenges of extending digital infrastructure across vast, sparsely populated regions (Kuersten 2018). This gap is further worsened by limited access to education and economic constraints, which hinder the ability of residents to afford and utilise technology effectively.

Concerning misinformation, it poses significant threats to sustainable urban development by eroding trust in public institutions, hindering public engagement, compromising decision-making, polarising communities, undermining climate action efforts, and affecting (mental) health and safety (Vasist and Krishnan 2023). It leads to scepticism towards sustainable initiatives and can stall or derail projects crucial for urban improvement. To combat its adverse effects, strategies such as enhancing digital literacy, prosecuting hate speech, robust fact-checking, transparent communication, community engagement, and cross-sector collaboration are essential for fostering informed public discourse and ensuring decisions are based on accurate information (Acuto *et al.* 2021).

Networks and Partnerships

The institutional set-ups and capacities of cities affecting policy-making at different governance levels also include the transnational activities of cities, which we will discuss in detail further below and also in the following parts of the book (see Chapter 6). City networks and partnerships are increasingly recognised as essential facilitators of sustainable urban development in a transnational context (Herrschel and Newman 2017). These collaborative arrangements can take various forms, such as networks of cities working together on common issues or partnerships between cities and other stakeholders such as businesses, non-profit organisations, or international agencies (Bansard *et al.* 2017).

Networks and partnerships enable cities to share experiences and best practices in sustainable urban development, fostering a learning environment through which cities can innovate and avoid costly mistakes (Busch 2016, Haupt *et al.* 2020). These collaborations enhance city capacities by providing technical assistance, training, and resources (Bouteligier 2013). Given that many sustainable development challenges, such as climate change and environmental pollution, transcend boundaries, city networks, and partnerships offer a vital platform for collaborative problem-solving and innovation (Dumała *et al.* 2021). Additionally, these networks can bolster the legitimacy of sustainability initiatives and collectively advocate for supportive policies and resources (Johnson 2018). By uniting, cities can influence national or international policy and mobilise resources for sustainable development (Lee 2019).

3.2.2 Actors and their Relationships

Sustainable urban development—understood as the process of planning and designing urban spaces to promote social equity, economic prosperity and environmental sustainability—is not just a technical or economic process but a profoundly political and social one. It is about negotiating different interests, making collective decisions and shaping public policy in ways that promote a more sustainable future (Dooling 2019). As a socio-political process, it involves various actors with different roles who influence decision-making at multiple levels, from local to national, shaping the future of urban environments and beyond (Bornemann *et al.* 2022). The process encompasses the following areas:

1. Policy-making and Regulation Setting: Elected officials and bureaucrats create political concepts and legal frameworks, such as zoning laws and environmental regulations, guiding urban development.
2. Political Advocacy: Interest groups influence policy through lobbying, public demonstrations or media campaigns, advocating for or against sustainable practices.
3. Public Participation: Citizens engage in development projects by expressing support or opposition, voicing concerns, and suggesting alternatives.
4. Negotiation and Compromise: The involvement of diverse stakeholders often necessitates negotiation and compromise.
5. Investment and Financing: Decisions by financial institutions and investors about where and what to invest in significantly affect urban sustainability.
6. Implementation and Enforcement: Various stakeholders, including city authorities, developers, and community groups, are responsible for implementing and enforcing policies and plans.
7. Education and Communication: Effective communication and education campaigns are vital for promoting understanding and acceptance of sustainable urban development among the community.

We consider the following as main actors (those who are directly affected by the issues, or those who have these responsibilities within the scope of their competence and have the will to act, in this sense, stakeholders) involved in the presented areas of policy- and decision-making processes shaping the sustainable urban development:

1. Residents, who are both the beneficiaries of urban development and should be key actors in decision-making;
2. Government entities (on different levels), who are both elected representatives and administration (particularly important are urban planners, architects, and engineers working in city hall);
3. NGOs and community-based organisations advocating for certain types of development, representing vulnerable communities, and sometimes directly implementing development projects;
4. Academia and independent experts providing research and knowledge that inform urban development practices, including studies on sustainable materials to sociological research on urban living; and
5. The private sector, including local real estate developers, construction companies, and infrastructure firms; moreover, banks and other financial institutions providing the financing needed for urban development like loans for construction projects.

Against this background, we propose the following set of critical roles that these actors perform:

Citizen/Voter: Engages directly in policymaking through public consultations, referenda, town hall meetings, electing officials, and exerting democratic control over urban development.

Policy-maker/decision-maker: Elected officials and appointed bureaucrats who set the strategic direction for city growth, establish policies and legislate on issues such as land use, building codes, and environmental standards.

Community organiser: Mobilises community residents to participate in urban development decisions, acting as a bridge between the community and policy-makers or developers.

Advocate: Lobbies for policies or projects, often representing interest groups such as community associations or environmental organisations.

Investor: Provides financial resources, both public and private, for urban development projects.

Developer: Designs and executes urban development projects in accordance with policies and market conditions, managing public opposition or support.

Regulator: Ensures compliance with policies, laws, and regulations in urban development.

Consultant/Advisor: Offers specialised advice and guidance in urban planning, architecture, or environmental science to stakeholders.

Researcher/Analyst/Expert: Studies urban development trends and provides data and expertise to inform policy and decision-making.

Communicator/media Professionals: Reports on urban development, explains policies, and shapes public opinion through various media channels.

The roles representing how individuals and organisations participate in and influence urban development policy-making can evolve and vary over time and context. These roles often interact and overlap, creating a dynamic and sometimes contentious political process. Dolmans *et al.* (2023) identify public leaders and managers as "conveners", "facilitators", and "catalysts", which are essential for fostering collaborative innovation and addressing challenges such as conflicting interests and uncertainties. This requires a shift in managerial operations, where politicians and leaders must collaborate with different groups and be willing to take risks, despite the potential for reputational damage from missteps (Dolmans *et al.* 2023).

3.2.3 Priorities for Sustainable Urban Development

Advancing the sustainable development of cities involves addressing various economic, social, and environmental priorities on different levels to ensure urban areas' long-term viability and resilience (Biermann *et al.* 2022). The following key priorities can be identified in policy documents—Urban Green Spaces and Green Building Practices, Public Sustainable Transportation, Resource Efficiency, Sustainable Waste Management Practices, Smart Technology, Affordable and

Sustainable Housing, and Health and Well-being (New Urban Agenda 2017, UN General Assembly 2015, Paris Agreement 2015)—and scholarly literature—Tang and Lee 2016, Watson 2016, Parnell 2016, Trindade *et al.* 2017, Reckien *et al.* 2017, Sanchez Rodriguez *et al.* 2018, Valencia *et al.* 2019, Caprotti *et al.* 2017, Song *et al.* 2023.

Urban Green Spaces and Green Building Practices

Urban Green Spaces (such as parks, gardens, protected areas and nature reserves, riverbanks, and boulevards) are perceived as an appropriate way to reduce urban heat island effects so common in southern urban areas and offer comfort to the nearby residents (Aram *et al.* 2019). They are dedicated to vegetation, nature and open spaces; as such, they are critical to maintaining high environmental quality, good human health, and enhancing the quality of life for residents in urban areas (Vargas-Hernández *et al.* 2023, Krellenberg *et al.* 2021). Their importance has, however, also led to the growing phenomenon of urban green grabbing[3] by residential real estate developers that are framed under the umbrella of sustainable development (García-Lamarca *et al.* 2022).

Green Building Practices, in turn, are construction, design, and localisation strategies that aim inter alia (1) to reduce the environmental impact of buildings (e.g., construction and operational waste reduction) and improve their material, energy, and water efficiency practices; (2) to minimise the carbon footprint and lead to long-term savings and improved health for building residents (via indoor environment's quality); (3) to enable easy access to public transportation, community resources; (4) to lower the costs of building operation and maintenance, which can include utilising intelligent building technologies for better control and optimisation of building performance; and (5) to preserve existing landscapes and ecosystems (Luo *et al.* 2022). Green Building Practices are often guided by various rating systems and certifications such as LEED (Leadership in Energy and Environmental Design), BREEAM (Building Research Establishment Environmental Assessment Method), and Green Star, which provide benchmarks for sustainability and environmental performance.

Public Sustainable Transportation

Public Sustainable Transportation involves developing and implementing public transit systems that embody principles of environmental responsibility, economic viability, and social equity (Kraus and Proff 2021). Sustainable public transportation systems are designed to be energy efficient and encourage the use of multiple modes of transport (Bamwesigye and Hlavackova 2019). Moreover, it is a sustainable practice as it reduces the number of private vehicles on the road, thereby reducing emissions per capita (Kraus and Proff 2021). Sustainable transportation considers environmental impacts, resilience, and adaptability and focuses on providing accessible and affordable transit options for all community members; this can be achieved through coordinating transportation planning with land use planning to encourage transit-oriented development. Particularly for cities in remote regions, Public Sustainable Transportation seems to be one of the leading challenges nowadays—insufficient transportation options have been identified as an obstacle to accessibility and social inclusion (Berg and Ihlström 2019).

Resource Efficiency

Resource efficiency is critical to sustainable urban development since it involves the use of limited resources and reduction of environmental impacts. Achieving resource efficiency and/for sustainable urban development relates to (1) investing in green infrastructure (managing stormwater, reducing heat island effect, and improving air quality); (2) developing sustainable modes of transport to reduce traffic, improve air quality, and promote health and well-being; (3) using renewable energy sources; (4) implementing energy-efficient technologies and optimising urban planning and building designs to reduce energy demand; (5) implementing efficient waste management systems, encouraging recycling, and composting and minimising waste production in the first place; and (6) implementing systems to reduce water consumption and improve water efficiency, such as rainwater harvesting, greywater recycling, and water-efficient appliances (Kabisch *et al.* 2018).

Sustainable Waste Management Practices

Cities that aim for sustainability develop an integrated approach to waste management that not only deals with waste after it has been generated but also prioritises waste reduction and recycling (Shukla and Hait 2022). Waste management refers to the activities and actions required to manage waste from its inception to its final disposal. Mainly, it involves the collection, transport, treatment, and disposal of waste, as well as its monitoring and regulation. It also encompasses the legal and regulatory framework for waste management and guidance on recycling and waste prevention (Khan *et al.* 2022).

Smart Technology

Smart technology plays a growing role in sustainable urban development because it can make cities more efficient, can improve the quality of life for residents, and can help urban areas become more sustainable and resilient (Duygan *et al.* 2022). The term "Smart City" (cf. Section 2.1) often refers to the integration of information and communication technology and the Internet of Things technology in a secure fashion to organise a city's assets, infrastructure, and processes (Toli and Murtagh 2020). Smart technology contributes to sustainable urban development through: (1) digital communication technology to detect and react to local changes in electricity usage (e.g., smart grids); (2) optimisation of traffic flow; (3) automatic systems that control the building's operations such as heating, lighting, security, and other systems; (4) advanced waste and water management systems, which optimise waste collection routes, reduce fuel consumption, predict waste generation patterns, track water usage in real-time, help detect leaks and provide data that can encourage more efficient water use; and (5) smart lighting which allows street lights to be dimmed or turned off when they are not needed (Richter *et al.* 2022). It should be highlighted that Smart Cities generate vast amounts of data that can be analysed to inform decision-making, predict trends and optimise urban services (Tura and Ojanen 2022). Moreover, the needed data storage is often quite energy-intense. The data can be particularly useful in urban planning, where data on traffic patterns, energy use, and other can help create more sustainable and efficient cities (Singh *et al.* 2022). Smart technologies can also enhance a city's ability to respond to and recover from disasters. This might involve monitoring systems that provide early warnings of floods and forest fires or digital platforms that improve emergency communication and coordination (Josipovic and Viergutz 2023). Moreover, digital media can be used to enhance communication between the city and its residents, allowing citizens to report problems, provide feedback and participate in decision-making processes (van Twist *et al.* 2023).

Affordable and Sustainable Housing

As urban populations usually grow, the demand for housing increases, leading to challenges such as urban sprawl, increased energy consumption, and rising housing costs. However, by integrating sustainability principles into housing development, cities can create liveable, inclusive, and sustainable urban environments (Moghayedi *et al.* 2021). A key aspect of sustainable housing is affordability, which can be achieved through efficient design and construction methods that reduce building costs and policies that ensure affordable housing is available to all sections of society. Sustainable housing also involves creating vibrant communities with easy access to amenities such as schools, parks, and public transportation (Moghayedi *et al.* 2021). Community engagement in the planning and design process can ensure that housing meets the needs and preferences of residents. By integrating these principles, cities can create affordable, sustainable housing that enhances the quality of life for residents, reduces environmental impacts, and contributes to social and economic sustainability (Wakely 2020). Policies and incentives can significantly promote sustainable and adequate to changing climate conditions housing and ensure it is accessible to all citizens (Trudeau 2018).

Health and Well-being

Sustainable cities promoting health and well-being can enhance citizens high quality of life, improve mental well-being, reduce healthcare costs, and create more resilient communities. This can be achieved through so-called "Healthy Urban Design", which involves designing cities for pedestrians and cyclists, with plenty of green spaces and access to amenities; well-planned urban design also includes ample public spaces that foster community interaction and social cohesion (Pineo *et al.* 2022). In turn, providing access to high-quality healthcare services is a fundamental aspect of urban health, which involves not only hospitals and clinics but also preventative and community health services, as was important during the COVID-19 pandemic (El Khateeb and Shawket 2022). Such an approach—"Health in All Policies"—involves considering the health impacts of policies across all sectors, not just health policy as such. To sum up, by prioritising the areas of health and well-being, cities can create environments that support their citizens' physical and mental health and contribute to broader social, economic, and environmental sustainability goals.

3.3 Drivers

3.3.1 Imaginaries

During the field research, our attention was drawn to a recurring theme often appearing in the interviewees' remarks: a particular sensitivity towards or awareness of the relationship between nature and urban space. Galland and Grønning call this a spatial consciousness and explain that it "may refer to the awareness about one's own spatial presence, as an individual or a community, and about real-world spatial phenomena and processes" (2019, p. 1). This relationship not only refers to the links between nature and the city in the context of sustainable urban development but also serves as a defining context for what the city and its surrounding nature mean to its citizens and how they define their territoriality (Gold 2019). Although most interviewees admitted that sustainable development is a relatively new phenomenon or policy, they simultaneously presented their personal or collective vision of space, geography, and environment in various ways and contexts. Nature and urban space—these two topics of sustainable urban development as a policy and collective visions seem so firmly and closely related that they merit a more detailed examination. We associate this observation, conceptually, with categories such as "geographical imaginaries", "spatial imaginaries", or "environmental imaginaries" (Watkins 2015, Howie and Lewis 2014, Peet and Watts 2004, Chhetri *et al.* 2023).

As Howie and Lewis argue, "(…) the idea of geographical 'imaginaries' is an attempt to capture not only that there are multiple geographical imaginations at large in the world, but that they do work in framing understandings of the world" (2014, p. 132). This concept enables us to observe and understand "popular, institutional, political and technical representations of the world as structured by more or less fixed and distinctive … framings of relations between people, place and territory" (Howie and Lewis 2014, p. 133). Although

> these framings may be intuited, discursive, textual or institutionalised, they shape and frame how people understand their worlds and those of others. … Such geographical imaginaries are constructions of the world, but are also 'vitally implicated' in the material and discursive making of the world.
>
> (Howie and Lewis 2014, p. 133)

Moreover, geographical imaginaries are immersed or interpenetrate with sociological imaginations, which are shaped by adopted systems of social values and identities (Harvey 2006, Chateau *et al.* 2021).

Chhetri *et al.*, in their study about linkages between geographical imaginaries and environmental governance, explain that (1) "there are always multiple imaginaries at play when it comes to the governance of the environment"; (2) "global-spanning imaginaries related to climate change, biodiversity, food security, and urbanisation get translated into local cultural contexts"; (3) "effective environmental governance requires imagination while also showing that imagination is not a panacea" (Chhetri *et al.* 2023, p. 263). Imaginaries of the geographical environment hence refer to how people perceive, understand, and conceptualise the geography and physical surroundings of an area or city (Howie and Lewis 2014), and they also influence place-making, in terms of design and using symbols (Huang and Roberts 2019). These ways encompass the range of cultural, historical, social, and personal understandings or frameworks that shape people's attitudes and behaviours towards geographical spaces. This can also include their collective urban aspirations (Bunnell 2019), fears and fantasies about specific geographical regions or features—sustainable urban development can be one such aspiration.

Based on our findings, we claim that imaginaries can impact sustainable urban development in the following ways. Firstly, on the local level, people's imaginaries often influence their level of engagement in community development initiatives. If a community perceives their environment as valuable, unique, or worth protecting, its members might be more likely to participate in sustainable practices or support sustainable development (Bunnell 2019). Conversely, seeing the environment as unimportant or expendable could lead to apathy or resistance to sustainable initiatives. Secondly, understanding or sharing the same imaginaries can help policy-makers and cities formulate development strategies or even specific urban and architectural designs that resonate with the local population and its needs.

Moreover, imaginaries might shape local expectations about what development should look like, which can significantly impact the success or failure of sustainable urban development projects. In addition, designs embedded in these local identities can incorporate sustainable practices, such as using local materials or integrating green spaces. Thirdly, understanding local imaginaries can be critical in developing strategies for climate change adaptation and building resilience because resilience in this context is also socially constructed. Communities that view their geographical surroundings as vulnerable might be more willing to invest in adaptive measures (as is shown in the case of Kolari; see Chapter 5). At the same time, those that perceive their environment as resilient may resist such efforts (as with Akureyri). Fourth, imaginaries can significantly influence the management of natural resources, such as, renewable energy. For example, if a community views a particular natural resource as sacred or significant, it might encourage more sustainable management of that resource. Fifth, imaginaries also contribute to developing specific discourses and labels which can be applied in city-branding strategies and development plans (e.g., Tromsø as a gateway to the Arctic, Kiruna as a mining city).

Likewise, urban development strategies and policies can also introduce new imaginaries to reshape a city's identity. And a city's imagined future can be used as a narrative to steer the city towards that future (King and Leandres 2019). For example, if citizens imagine the city as a tech hub, urban development strategies might prioritise the development of tech industries, innovation hubs, and related infrastructure (as in the case of Luleå). Finally, imaginaries of Indigenous peoples living in the region where cities pursue sustainable urban development often challenge or even clash with the imaginaries of non-Indigenous residents. The implication of this situation is related to land rights and land use traditions, geo-cultural preservation, and the inclusion of Indigenous peoples' imaginaries into local policies based on cultural sensitivity and respect for the challenges that come with climate change. Consequently, we claim that imaginaries provide different visions of what cities can be, reflecting diverse priorities and values. While they are powerful tools for urban planning and policy-making, they must also be critically examined to consider whose interests they serve and how they might impact sustainability outcomes. Finally, we also expect that imaginaries play an important role in envisioning, designing, and framing policy goals on the others levels of governance; however, this question remains open for further studies (Liverman 2018, Immler and Sakkers 2022).

3.3.2 Cooperation

In this part, we focus on the transnational dimension of sustainable urban development as the area where aligning the global ambitions and local needs could be achieved. At the same time, we primarily aim to identify causes that support or limit the ability of cities to cooperate with different types of actors at or beyond the local level.

In the realm of international relations, cities' influence and roles have been in flux across the ages.[4] However, after the 1648 Treaties of Westphalia, nation-states became the predominant international players for many years. This dynamic began to shift in the latter half of the 20th century. New global economic and political conditions enabled non-state entities to play active roles in global affairs (Curtis and Acuto 2018). Today's international relations represent a complex interplay involving various actors, including states and cities. The latter engage diversely based on specific issues, interests, and capacities within a multifaceted diplomatic landscape, resulting in a redistribution of responsibilities and resources (van der Pluijm and Melissen 2007). As we moved into the 21st century, city diplomacy has resurfaced, defined as the ways cities or local governments interact internationally to represent and protect their interests against external influences (van der Pluijm and Melissen 2007, p. 6). Some, like Barber (Barber 2013), argue that cities and their leaders are better positioned than nation-states to tackle global challenges, due to their pragmatism, inclusive approach and innovative spirit. Many cities are inclined to address intricate challenges using trust, public involvement, disregard for borders, and innovative solutions.

As we mentioned, transnational city collaboration is a concept that refers to cities across different nations working together to address common challenges or pursue shared goals. This collaboration can be related to other fields, take many forms and use various methods and tools (an overview of this variety is presented

Table 3.1 (Transnational) Cooperation between cities: fields, forms, methods and tools

Fields	Economic cooperation; Policy alignment; Technological cooperation; Cultural exchange; Environmental sustainability; Security and crisis management; Infrastructure development; Education and research collaboration.
Forms	Twin (sister) City relationships; City networks; participation in formal and direct forms of international cooperation (e.g., Nordic, in EU, climate neutral cities network); Public-private partnerships; Direct bilateral agreements; Exchange of knowledge and best practices; Multilateral projects or initiatives; Joint ventures; Collaborative Research and development (R&D); Exchange Programmes and Visits; Inter-city Competitions; Shared governance models.
Methods and tools	Online platforms and databases; Virtual and physical meetings/ Conferences; Collaboration software; Social media and websites; Legal agreements; Common standards and frameworks; Administrative technical tools; Joint funding mechanisms; Training and capacity building programmes; Metrics and indicators; Common standards and benchmarking

Source: Authors compilation based on Acuto 2013, Bouteligier 2013, Curtis 2014, Bansard *et al.* 2017, Curtis and Acuto 2018, Johnson 2018, Lee and Jung 2018, Oosterlynck *et al.* 2018, Davidson *et al.* 2019, Haupt *et al.* 2020, Gordon 2020, Lee *et al.* 2020, Sassen 2002, Szpak *et al.* 2022.

in Table 3.1). It often centres on issues that transcend city boundaries and require collective action, such as climate change, economic development, social equity, and urban sustainability. The increasing prevalence of transnational city collaboration reflects a shift in how we understand and approach global issues. While nation-states continue to play a vital role in global governance, cities are increasingly recognised as critical actors in addressing global challenges. This is partly due to their proximity to citizens and daily issues as well as to extraordinary situations (such as the management of the COVID-19 pandemic or welcoming refugees), their economic power and their often-innovative and pragmatic, hands-on approach to problem-solving.

City diplomacy is on the rise, yet cities remain reliant on higher levels of government—usually in terms of legal framework, fiscal issue, or political pressures. In their interactions with national and supranational bodies, cities often find themselves in a less advantageous position due to inherent power imbalances, essentially seeking favours or even permission from superior authorities (Pipa and Bouchet 2020). On a global stage, city diplomacy can be bilateral, such as sister city/twin city relationships, or multilateral, involving various city representatives in groups such as municipal networks (van der Pluijm and Melissen 2007). These diplomatic endeavours span a range of topics, from security and economic development to educational and cultural exchanges.

Increasingly, the goal of city diplomacy is to have a voice in and impact decision-making in broader platforms such as the EU or even globally. Such efforts are bolstered by activities such as international networking and collaborative institution-building. These networks are geared towards both representing city interests in various decision-making forums and facilitating the exchange of best practices on city-specific issues. Notable networks include the World Organisation of United Cities and Local Governments (UCLG) and regional bodies such as the Council of European Municipalities and Regions. Other significant groups include ICLEI Local Governments for Sustainability, the Climate Alliance, C40 Cities Climate Leadership Group, the Global Compact of Mayors, and the Carbon Neutral Cities Alliance. Scholars highlight that such transnational city networks make bold propositions and enhance policy learning, potentially navigating the limitations set by regional or national frameworks (Fuhr *et al.* 2018). Moreover, city networks steadily pool resources, set priorities, share strategies, and document emissions cuts, thereby enhancing collaboration and communication between their members (Gordon and Johnson 2018).

Today, city alliances are significant players in global governance (Kraas *et al.* 2016). Much like other networks, they have gained a reputation as influential entities or even frontrunners in global diplomacy (Acuto 2013). Some experts, such as van der Pluijm and Melissen (2007), believe that these networks are gradually taking the place of nation-states in diplomacy, with cities playing a pivotal role in this shift. Pipa and Bouchet (2020) further stress that, for cities to secure their positions in global decision-making forums, they must consistently champion their collective interests. This perspective encompasses both regional and global city alliances, including The Global Parliament of Mayors, Metropolis, the C40 Cities Climate Leadership Group, and the Global Covenant of Mayors (Bouteligier 2013). Currently, there are roughly 300 such transnational city networks, and this number is on the rise.

Cities are playing a key role in achieving the goals of the Paris Climate Agreement and the Sustainable Development Goals (SDGs) for several reasons. Firstly, over half of the global population resides in cities, a figure projected to rise to two-thirds by 2050 (Kraas *et al.* 2016). Secondly, cities account for over 70% of global CO_2 emissions. Lastly, cities are viewed as hubs of growth, innovation, and wealth (António Guterres, UN Secretary-General, cited Coalition for Urban Transitions 2019). They are seen as transformative players in crucial areas of climate and sustainable development, notably urban land use, materials flow, and urban health (Kraas *et al.* 2016). Consequently, city administrations are anticipated to harmonise their strategies with climate goals, addressing two primary global concerns: enhancing the quality of life and curbing climate change.

In regard to the Paris Climate Agreement, for instance, city alliances, along with the collaboration of other subnational entities in International Cooperative Initiatives (ICIs), are seen as avenues to support national governments in fulfilling their Nationally Determined Contributions (NDCs) by implementing bold measures that exceed national policies (NewClimate Institute *et al.* 2019). But what are the practical implementations and outcomes of such endeavours since climate governance and sustainable development, cities, regions, and businesses are poised to ensure that global climate strategies bolster local sustainable endeavours instead of impeding them (NewClimate Institute *et al.* 2019)? Ideally, approaches aiming to foster the SDGs and the Paris Agreement would involve cities and local administrations as these are the ground where global agreements are actualised (Fenton and Gustafsson 2017). Cities symbolise areas where individuals can envision an enhanced quality of life in a renewed urban framework. The COVID-19 crisis underscored the significance of cities in the face of a needed immediate crisis response, given their ability to manage local challenges while upholding regional and global commitments. For instance, cities showcased their climate initiatives on the "Global Climate Action Portal" prior to the Paris Climate Agreement's ratification in 2015.

In global climate oversight, cities and their alliances are seen as potential facilitators, aiding the amplification of local climate initiatives (Fuhr *et al.* 2018). Discussions frequently pivot towards orchestration and policy diffusion in the relevant research. Gordon and Johnson reference studies that demonstrate how city networks guide member cities by disseminating norms, methodologies and concepts (Gordon and Johnson 2018). Yet, scholars highlight the multiple challenges cities confront, especially when collaborating in alliances targeting climate objectives and SDGs. While local governments' roles vary regionally, common challenges include a limited jurisdiction in pivotal climate sectors and scarce resources for persistent climate action (Fuhr *et al.* 2018), alongside a dependency on other governmental tiers. Concerning city alliances, it is vital to acknowledge their position amidst the political dynamics influencing their efficacy (Gordon and Johnson 2018). Therefore, to grasp the capabilities of city alliances, it is essential to examine diverse real-world cases and identify their nuances.

At this point, the issue of causes that support or limit the activity of cities in such networks, or, looking a bit more broadly, in the transnational space, arises. Following the research presented by Mokhles and Davidson (2021), we distinguish several determinants which shape involvement of municipality in city-networks, including:

- geographical location and other geographic features (distance to a neighbouring state, specific landscape—for example, island or belonging to a specific region, for example, transboundary historical region or seismic-prone region can serve as platform for shared interests and stimulate exchanges between cities (Hansen and Coenen 2015);
- accessibility and transport connectivity, since, despite growing virtual connections, they still play a role in city networking by municipalities located in remote regions (Navarro-Azorín *et al.* 2022);
- international recognition and reputation which can be included as elements considered by cities while engaging into closer collaboration (Nielsen and Papin 2021); this issue is also related with city-branding strategies (Bonakdar and Audirac 2020);
- well-developed or advanced sectors, including economy, research, and culture, which are open for international contacts, exchanges, or inflows are another source of involvement of cities in transnational cooperation (Levent-Baycan *et al.* 2008);

hosting international institutions or events (Szpak *et al.* 2022);
demographic structure offering openness and people-to-people contacts abroad, for example, presence of language or cultural minorities, diaspora communities, and immigrants (Head *et al.* 2019); and
political will and strategies (Prado-Lorenzo *et al.* 2012).

While the list of determinants is not exhaustive, we consider those above as crucial for cooperation in the context of sustainable urban development.

3.4 Conclusions

In this section, we established the conceptual framework for our further investigation, which is a model composed of three factors and two drivers. It is an original framework that integrates existing knowledge with findings from our field works and analysis.

We identified and discussed in five parts the essential elements determining the content and the shape of the policies and solutions adopted for sustainable urban development in remote regions. We discussed the institutional set-ups and capacities affecting policy-making for sustainable urban development at different governance levels. Next, we investigated the roles assigned to actors involved in urban development processes. In the next part, we categorised the priorities of sustainable urban policies, which have guided cities worldwide in recent decades. Afterwards, we discussed how imaginaries shape people(s)' approaches and institutions' policies towards sustainability in their cities. Finally, we focused on the transnational dimension of sustainable urban development as the area where aligning the global ambitious and local needs could be achieved.

Our analytical framework was developed to help to detect and understand: (1) why local approaches to sustainable urban development differ; (2) how complex urban governance arrangements deal with the entanglements of socioeconomic, political, and environmental processes shaping urban development; (3) how different actors engage in the policy-making and what difficulties they face on this way; and, finally, (4) how cooperation can stimulate sustainable urban development in the remote regions.

The main advantage of this model is threefold. First, it enables analysing how complex policy- and decision-making is combined in a particular space and time; this combination produces certain (but not unique) decisions and outcomes based on different political and legal settings, diversified capacities and sources of knowledge(s). Second, it allows us to compare approaches and policies with other cities located in remote regions since it can suggest some indicators, parameters related with some of the key elements. Third, pointing out barriers and obstacles provides lessons on how sustainable urban development can be improved through better alignment between global aspirations and local possibilities and needs and how to make it more inclusive based on participatory processes. The limitations of the proposed model are discussed in the last chapter. Finishing up, although Baud *et al.* (2021) claim that such models are always insufficient in the face of complex challenges and circumstances, we still believe that they can provide useful framework for research and better understanding these complexities.

Acknowledgements

This chapter was written by Michał Łuszczuk. The outline was developed by Michał Łuszczuk and Dorothea Wehrmann. The graphic was designed by Katharina Schaarschmidt (IDOS). The chapter was reviewed by Jacqueline Götze and Arne Riedel. The chapter was further reviewed by an external expert. An earlier version of this chapter was discussed at the "Urban Issues" workshop at IDOS in August 2023. All authors of this book have read and accept the content of this chapter.

Notes

1 These frames help state and non-state actors alike organise their thoughts, engage in meaningful dialogue, and make decisions that align with the overarching goals of sustainability. Some key conceptual frames of SD include: Triple Bottom Line (Alhaddi 2015), Ecological Footprint (Venetoulis and Talberth 2008), Circular Economy (Schroeder *et al.* 2019), or Planetary Boundaries (Randers *et al.* 2019). Understanding these conceptual frames is crucial for policy-makers, businesses, and individuals as they navigate the complexities of sustainable development. By applying these frames to different challenges, state and non-state actors can derive strategies and solutions that align with long-term sustainability goals (Soergel *et al.* 2021, Jong and Vijge 2021, Soergel *et al.* 2021, Jong i Vijge 2021).

2 The Arctic Urban Database is a result of the Arctic Partnerships in International Research and Education (PIRE) research collaboration, a National Science Foundation-funded project focused on promoting urban sustainability in the Arctic. Available from https://www.arcticurbandata.org/ (Accessed 08 August 2023).

3 Urban green grabbing refers to the appropriation of urban green spaces by powerful entities—such as governments, developers, or corporations—under the guise of environmental or sustainable development. This practice often leads to the displacement of local communities, the reduction in public access to green areas, and the potential for social injustices, despite being presented as beneficial for urban sustainability or green infrastructure. For more see: García-Lamarca *et al.* (2022).

4 This has been influenced by the cities' internal changes, mainly shifts in their potential and abilities, and due to overarching global structural changes and societal evolution (Oosterlynck *et al.* 2018, Curtis 2014). Historical events reflect highs and lows of ancient Greek city-states and the rise and fall of influential medieval European cities. Collaborations among these cities took various shapes, such as ancient Greek federations including the Delian League, or alliances in medieval Europe such as the Hanseatic League. Additionally, the dominant Italian city-states pioneered the concept of permanent diplomatic missions.

References

Acuto, M., 2013. City leadership in global governance. *Global Governance*, 19 (3), 481–498.

Acuto, M., *et al.*, 2021. Mobilising urban knowledge in an infodemic: Urban observatories, sustainable development and the COVID-19 crisis. *World Development*, 140, 105295.

Ahad, M. A., Casalino, G., and Bhushan, B., 2023. *Enabling technologies for effective planning and management in sustainable smart cities*. Cham: Springer.

Alhaddi, H., 2015. Triple bottom line and sustainability: A literature review. *Business and Management Studies*, 1 (2), 6.

Allain-Dupré, D., 2020. The multi-level governance imperative. *The British Journal of Politics and International Relations*, 22 (4), 800–808.

Aram, F., *et al.*, 2019. Urban green space cooling effect in cities. *Heliyon*, 5 (4), e01339.

Arctic Urban Database 2023. ARCTIC URBAN DATABASE [online]. Available from: https://www.arcticurbandata.org/ [Accessed 8 August 2023].

Åström, J., 2019. Citizen participation. *In:* Orum, A. M., ed., *The Wiley Blackwell encyclopedia of urban and regional studies*. Chichester: Wiley, 1–4.

Bamwesigye, D., and Hlavackova, P., 2019. Analysis of sustainable transport for smart cities. *Sustainability*, 11 (7), 2140.

Bansard, J. S., Pattberg, P. H., and Widerberg, O., 2017. Cities to the rescue? Assessing the performance of transnational municipal networks in global climate governance. *International Environmental Agreements: Politics, Law and Economics*, 17 (2), 229–246.

Barber, B. R., 2013. *If* may*ors ruled the world. Dysfunctional nations, rising cities*. New Haven, CT: Yale University Press.

Bardal, K. G., *et al.*, 2021. Factors facilitating the implementation of the sustainable development goals in regional and local planning—Experiences from Norway. *Sustainability*, 13 (8), 4282.

Baud, I., *et al.*, 2014. *Participatory "spatial" knowledge management configurations in metropolitan governance networks for SD*. (Chance2Sustain thematic reports; No. 2). EADI/Chance2Sustain. http://www.chance2sustain.eu/fileadmin/Website/Dokumente/Dokumente/Publications/publications_2014/C2S_TR_No02_WP5_V5-6.pdf.

Baud, I., *et al.*, 2021. The urban governance configuration: A conceptual framework for understanding complexity and enhancing transitions to greater sustainability in cities. *Geography Compass*, 15 (5), e12562. https://doi.org/10.1111/gec3.12562.

Berg, J., and Ihlström, J., 2019. The importance of public transport for mobility and everyday activities among rural residents. *Social Sciences*, 8 (2), 58.

Berman, T., 2017. *Public participation as a tool for integrating local knowledge into spatial planning*. Cham: Springer International Publishing.

Biermann, F., Hickmann, T., and Sénit, C.- A., eds., 2022. *The political impact of the sustainable development goals*. New York: Cambridge University Press.

Bisogno, M., *et al.*, 2023. Sustainable development goals in public administrations: Enabling conditions in local governments. *International Review of Administrative Sciences*, 89(4), 1223–1242.

Bonakdar, A., and Audirac, I., 2020. City branding and the link to urban planning: Theories, practices, and challenges. *Journal of Planning Literature*, 35 (2), 147–160.

Bornemann, B., Knappe, H., and Nanz, P., eds., 2022. *The Routledge handbook of democracy and sustainability*. Abingdon, New York: Routledge.

Bossuyt, D. M., and Savini, F., 2018. Urban sustainability and political parties: Eco-development in Stockholm and Amsterdam. *Environment and Planning C: Politics and Space*, 36 (6), 1006–1026.

Bouteligier, S., 2013. *Cities, networks, and global environmental governance. Spaces of innovation, places of leadership*. New York: Routledge.

Bouzguenda, I., Alalouch, C., and Fava, N., 2019. Towards smart sustainable cities: A review of the role digital citizen participation could play in advancing social sustainability. *Sustainable Cities and Society*, 50, 101627.

Bramwell, A., 2020. Innovation and the "ordinary" city? Urban policy making in a digital age. *PS: Political Science & Politics*, 53 (1), 15–19.

Bulkeley, H., and Betsill, M., 2005. Rethinking sustainable cities: Multilevel governance and the "urban" politics of climate change. *Environmental Politics*, 14 (1), 42–63.

Bulkeley, H., and Newell, P., 2010. *Governing climate change*. London: Routledge.

Bunnell, T., 2019. Urban aspirations. *In:* Orum, A. M., ed., *The Wiley Blackwell encyclopedia of urban and regional studies*. Chichester: Wiley, 1–3.

Busch, H., 2016. *Entangled cities transnational municipal climate networks and urban governance*. Lund: Lund University.

Camagni, R., 2017. Sustainable urban development: Definition and reasons for a research programme. *In:* Capello, R., ed., *Seminal studies in regional and erban economics*. Cham: Springer International Publishing, 261–282.

Caprotti, F., *et al.*, 2017. The new urban agenda: Key opportunities and challenges for policy and practice. *Urban Research & Practice*, 10 (3), 367–378.

Chateau, Z., Devine-Wright, P., and Wills, J., 2021. Integrating sociotechnical and spatial imaginaries in researching energy futures. *Energy Research & Social Science*, 80, 102207.

Cheshmehzangi, A., and Dawodu, A., 2019. *Sustainable urban development in the age of climate change*. Singapore: Springer Singapore.

Chhetri, N., Ghimire, R., and Eisenhauer, D. C., 2023. Geographies of imaginaries and environmental governance. *The Professional Geographer*, 75 (2), 263–268.

Coalition for Urban Transitions 2019. *The New Urban Opportunity.*

Curtis, S., ed., 2014. *The power of cities in international relations*. New York: Routledge.

Curtis, S., and Acuto, M., 2018. The foreign policy of cities. *The RUSI Journal*, 163 (6), 8–17.

Das-Doyle, C., *et al.* (2023), "The state of play and prospects for measuring innovation in the public sector", OECD Working Papers on Public Governance, No. 67, OECD Publishing, Paris, https://doi.org/10.1787/dca76af0-en

Davidson, K., Coenen, L., and Gleeson, B., 2019. A decade of C40: Research insights and agendas for city networks. *Global Policy*, 10 (4), 697–708.

Dolmans, S. A. M., *et al.*, 2023. A dynamic perspective on collaborative innovation for smart city development: The role of uncertainty, governance, and institutional logics. *Organization Studies*, 44 (10), 1577–1601.

Dooling, S., 2019. Urban sustainability. *In:* Orum, A. M., ed., *The Wiley Blackwell encyclopedia of urban and regional studies*. Chichester: Wiley, 1–13.

Dumała, H., *et al.*, 2021. Transnational municipal networks as a mechanism for marine governance toward climate change adaptation and mitigation: Between potential and practice. *Frontiers in Marine Science*, 8, 1–16, https://doi.org/10.3389/fmars.2021.626119.

Duygan, M., *et al.*, 2022. Where do smart cities grow? The spatial and socio-economic configurations of smart city development. *Sustainable Cities and Society*, 77, 103578.

Ehnert, F., Egermann, M., and Betsch, A., 2022. The role of niche and regime intermediaries in building partnerships for urban transitions towards sustainability. *Journal of Environmental Policy & Planning*, 24 (2), 137–159.

El Khateeb, S., and Shawket, I. M., 2022. A new perception; generating well-being urban public spaces after the era of pandemics. *Developments in the Built Environment*, 9, 100065.

Fenton, P., and Gustafsson, S., 2017. Moving from high-level words to local action—governance for urban sustainability in municipalities. *Current Opinion in Environmental Sustainability*, 26–27, 129–133.

Fuhr, H., Hickmann, T., and Kern, K., 2018. The role of cities in multi-level climate governance: local climate policies and the 1.5°C target. *Current Opinion in Environmental Sustainability*, 30, 1–6.

Galland, D., and Grønning, M., 2019. Spatial consciousness. *In:* Orum, A. M., ed., *The Wiley Blackwell encyclopedia of urban and regional studies*. Chichester: Wiley, 1–9.

García-Lamarca, M., *et al.*, 2022. Urban green grabbing: Residential real estate developers discourse and practice in gentrifying Global North neighborhoods. *Geoforum*, 128, 1–10.

Gold, J. R., 2019. Territoriality. *In:* Orum, A. M., ed., *The Wiley Blackwell encyclopedia of urban and regional studies*. Chichester: Wiley, 1–10.

Gonzales-Iwanciw, J., Dewulf, A., and Karlsson-Vinkhuyzen, S., 2020. Learning in multi-level governance of adaptation to climate change – a literature review. *Journal of Environmental Planning and Management*, 63 (5), 779–797.

Gordon, D. J., 2020. *Cities on the world stage*. Cambridge: Cambridge University Press.

Gordon, D. J., and Johnson, C. A., 2018. City-networks, global climate governance, and the road to 1.5°C. *Current Opinion in Environmental Sustainability*, 30, 35–41.

Hansen, T., and Coenen, L., 2015. The geography of sustainability transitions: Review, synthesis and reflections on an emergent research field. *Environmental Innovation and Societal Transitions*, 17, 92–109.

Harvey, D., 2006. The sociological and geographical imaginations. *International Journal of Politics, Culture, and Society*, 18 (3–4), 211–255.

Haupt, W., *et al.*, 2020. City-to-city learning within climate city networks: definition, significance, and challenges from a global perspective. *International Journal of Urban Sustainable Development*, 12 (2), 143–159.

Head, L., Klocker, N., and Aguirre-Bielschowsky, I., 2019. Environmental values, knowledge and behaviour: Contributions of an emergent literature on the role of ethnicity and migration. *Progress in Human Geography*, 43 (3), 397–415.

Herrschel, T., and Newman, P., eds., 2017. *Cities as international actors. Urban and regional governance beyond the nation state*. London: Palgrave Macmillan UK.

Howie, B., and Lewis, N., 2014. Geographical imaginaries: Articulating the values of geography. *New Zealand Geographer*, 70 (2), 131–139.

Huang, S.-M., and Roberts, J. L., 2019. Place-making. *In:* Orum, A. M., ed., *The Wiley Blackwell encyclopedia of urban and regional studies*. Chichester: Wiley, 1–5.

Immler, N. L., and Sakkers, H., 2022. The UN-Sustainable Development Goals going local: Learning from localising human rights. *The International Journal of Human Rights*, 26 (2), 262–284.

James, P., *et al.*, 2015. *Urban sustainability in theory and practice. Circles of sustainability*. London, New York: Routledge.

Johnson, C. A., 2018. *The power of cities in global climate politics*. London: Palgrave Macmillan UK.

Jong, E. de, and Vijge, M. J., 2021. From millennium to sustainable development goals: Evolving discourses and their reflection in policy coherence for development. *Earth System Governance*, 7, 100087.

Josipovic, N., and Viergutz, K., 2023. Smart solutions for municipal flood management: Overview of literature, trends, and applications in German cities. *Smart Cities*, 6 (2), 944–964.

Kabisch, S., *et al.*, eds., 2018. *Urban transformations*. Cham: Springer International Publishing.

Kagan, S., *et al.*, 2018. Culture in sustainable urban development: Practices and policies for spaces of possibility and institutional innovations. *City, Culture and Society*, 13, 32–45.

Keuffer, N., and Mabillard, V., 2020. Administrative openness and diversity in Swiss municipalities: how does local autonomy influence transparency practices? *International Review of Administrative Sciences*, 86 (4), 782–798.

Khan, A. H., *et al.*, 2022. Current solid waste management strategies and energy recovery in developing countries - State of art review. *Chemosphere*, 291 (Pt 3), 133088.

King, C. R., and Leandres, S., 2019. Image of the city. *In:* Orum, A. M., ed., *The Wiley Blackwell encyclopedia of urban and regional studies*. Chichester: Wiley, 1–5.

Kolehmainen, J., *et al.*, 2016. Quadruple helix, innovation and the knowledge-based development: Lessons from remote, rural and less-favoured regions. *Journal of the Knowledge Economy*, 7 (1), 23–42.

Kraas, F., *et al.*, 2016. *Humanity on the move: Unlocking the transformative power of cities. Flagship report*. Berlin: Wissenschaftlicher Beirat der Bundesregierung Globale Umweltveränderungen.

Kraus, L., and Proff, H., 2021. Sustainable urban transportation criteria and measurement—A systematic literature review. *Sustainability*, 13 (13), 7113.

Krellenberg, K., *et al.*, 2021. What to do in, and what to expect from, urban green spaces – Indicator-based approach to assess cultural ecosystem services. *Urban Forestry & Urban Greening*, 59, 126986.

Kuersten, A., 2018. The Arctic digital divide. *In:* O'Donnell, B., Gruenig, M., and Riedel, A., eds., *Arctic summer college yearbook*. Cham: Springer International Publishing, 93–105.

Ladner, A., *et al.*, 2019. *Patterns of local autonomy in Europe*. Cham: Springer International Publishing.

Lee, T., 2019. Network comparison of socialization, learning and collaboration in the C40 cities climate group. *Journal of Environmental Policy & Planning*, 21 (1), 104–115.

Lee, T., and Jung, H. Y., 2018. Mapping city-to-city networks for climate change action: Geographic bases, link modalities, functions, and activity. *Journal of Cleaner Production*, 182, 96–104.

Lee, T., Yang, H., and Blok, A., 2020. Does mitigation shape adaptation? The urban climate mitigation-adaptation nexus. *Climate Policy*, 20 (3), 341–353.

Levent-Baycan, T., Kundak, S., and Gulumser, A. A., 2008. City-to-city linkages in a mobile society: the role of urban networks in Eurocities and Sister Cities. *International Journal of Services Technology and Management*, 10 (1), 83.

Lissandrello, E., *et al.*, 2023. *The "new normal" in planning, governance and participation*. Cham: Springer International Publishing.

Liverman, D. M., 2018. Geographic perspectives on development goals. *Dialogues in Human Geography*, 8 (2), 168–185.

Luo, W., *et al.*, 2022. A systematic review of green construction research using scientometrics methods. *Journal of Cleaner Production*, 366, 132710.

Manzi, T., *et al.*, eds., 2015. *Social sustainability in urban areas. Communities, connectivity and the urban fabric*. London: Routledge.

Medeiros, E., and van der Zwet, A., 2020. Sustainable and integrated urban planning and governance in metropolitan and medium-sized cities. *Sustainability*, 12 (15), 5976.

Mersal, A., 2016. Sustainable urban futures: Environmental planning for sustainable urban development. *Procedia Environmental Sciences*, 34, 49–61.

Moghayedi, A., *et al.*, 2021. A critical success factor framework for implementing sustainable innovative and affordable housing: A systematic review and bibliometric analysis. *Buildings*, 11 (8), 317.

Mokhles, S., and Davidson, K., 2021. A framework for understanding the key drivers of cities' climate actions in city networks. *Urban Climate*, 38, 100902.

Navarro-Azorín, J. M., Artal-Tur, A., and Ramos-Parreño, J. M., 2022. Geography and embeddedness in city networks. *Spatial Economic Analysis*, 17 (2), 206–222.

NewClimate Institute, *et al.*, 2019. *Global Climate Action from Cities Regions and Businesses*. NA. Cologne, Germany: NewClimate Institute.

New Urban Agenda 2017. Nairobi: United Nations. Available from: https://habitat3.org/wp-content/uploads/NUA-English.pdf [17 May 2024].

Nielsen, A. B., and Papin, M., 2021. The hybrid governance of environmental transnational municipal networks: Lessons from 100 Resilient Cities. *Environment and Planning C: Politics and Space*, 39 (4), 667–685.

Oosterlynck, S., *et al.*, eds., 2018. *The city as a global political actor*. Abingdon, New York: Routledge.

Pandit, A., *et al.*, 2017. Infrastructure ecology: an evolving paradigm for sustainable urban development. *Journal of Cleaner Production*, 163, S19–S27.

Paris Agreement 2015. *Paris Agreement.* Available from: https://wedocs.unep.org/20.500.11822/20830 [17 May 2024.

Parnell, S., 2016. Defining a global urban development agenda. *World Development*, 78, 529–540.

Peet, R., and Watts, M., eds., 2004. *Liberation ecologies. Environment, development, social movements*. London: Routledge.

Petrov, A. N., 2017. Human capital and sustainable development in the Arctic: Towards intellectual and empirical framing. *In:* Fondahl, G., and Wilson, G. N., eds., *Northern sustainabilities: Understanding and addressing change in the circumpolar world.* Cham: Springer International Publishing, 203–220.

Pfeffer, K., *et al.*, 2013. Participatory spatial knowledge management tools. *Information, Communication & Society*, 16 (2), 258–285.

Pineo, H., Moore, G., and Braithwaite, I., 2022. Incorporating practitioner knowledge to test and improve a new conceptual framework for healthy urban design and planning. *Cities & Health*, 6 (5), 906–921.

Pipa, A. F., and Bouchet, M., 2020. Multilateralism restored? City diplomacy in the COVID-19 era. *The Hague Journal of Diplomacy*, 15 (4), 599–610.

Prado-Lorenzo, J.-M., García-Sánchez, I.-M., and Cuadrado-Ballesteros, B., 2012. Sustainable cities: Do political factors determine the quality of life? *Journal of Cleaner Production*, 21 (1), 34–44.

Randers, J., *et al.*, 2019. Achieving the 17 sustainable development goals within 9 planetary boundaries. *Global Sustainability*, 2, 1–11.

Rasoolimanesh, S. M., *et al.*, 2019. How governance influences the components of sustainable urban development? *Journal of Cleaner Production*, 238, 117983.

Reckien, D., *et al.*, 2017. Climate change, equity and the sustainable development goals: An urban perspective. *Environment and Urbanization*, 29 (1), 159–182.

Richter, M. A., *et al.*, 2022. Smart cities, urban mobility and autonomous vehicles: How different cities needs different sustainable investment strategies. *Technological Forecasting and Social Change*, 184, 121857.

Ruiz-Mallén, I., March, H., and Satorras, M., 2022. *Urban resilience to the climate emergency*. Cham: Springer International Publishing.

Russel, D., and Kirsop-Taylor, N., eds., 2022. *Handbook on the governance of sustainable development*. Cheltenham: Edward Elgar Publishing.

Salet, W., and Vries, J. de, 2019. Contextualisation of policy and law in sustainable urban development. *Journal of Environmental Planning and Management*, 62 (2), 189–204.

Sanchez Rodriguez, R., Ürge-Vorsatz, D., and Barau, A. S., 2018. Sustainable development goals and climate change adaptation in cities. *Nature Climate Change*, 8 (3), 181–183.

Sassen, S., 2002. *Global networks, linked cities*. New York: Routledge.

Schroeder, P., Anggraeni, K., and Weber, U., 2019. The relevance of circular economy practices to the sustainable development goals. *Journal of Industrial Ecology*, 23 (1), 77–95.

Shukla, S., and Hait, S., 2022. Smart waste management practices in smart cities: Current trends and future perspectives. *In:* Hussain, Ch., and Hait, S., eds., *Advanced organic waste management*. Amsterdam: Elsevier, 407–424.

Singh, T., *et al.*, 2022. A decade review on smart cities: Paradigms, challenges and opportunities. *IEEE Access*, 10, 68319–68364.

Sodiq, A., *et al.*, 2019. Towards modern sustainable cities: Review of sustainability principles and trends. *Journal of Cleaner Production*, 227, 972–1001.

Soergel, B., *et al.*, 2021. A sustainable development pathway for climate action within the UN 2030 agenda. *Nature Climate Change*, 11 (8), 656–664.

Sofyani, H., Riyadh, H. A., and Fahlevi, H., 2020. Improving service quality, accountability and transparency of local government: The intervening role of information technology governance. *Cogent Business & Management*, 7 (1), 1735690.

Song, K., *et al.*, 2023. Urban governance: A review of intellectual structure and topic evolution. *Urban Governance,* 3(3), 169–185.

Szpak, A., Modrzyńska, J., and Dahl, M., 2022. European cities, international relations and some popular connotations. *European Planning Studies*, 30 (6), 1034–1054.

Tang, H.-T., and Lee, Y.-M., 2016. The making of sustainable urban development: A synthesis framework. *Sustainability*, 8 (5), 492.

Toli, A. M., and Murtagh, N., 2020. The concept of sustainability in smart city definitions. *Frontiers in Built Environment*, 6, 1–10.

Trindade, E. P., *et al.*, 2017. Sustainable development of smart cities: A systematic review of the literature. *Journal of Open Innovation: Technology, Market, and Complexity*, 3 (3), 1–14.

Trudeau, D., 2018. Integrating social equity in sustainable development practice: Institutional commitments and patient capital. *Sustainable Cities and Society*, 41, 601–610.

Tura, N., and Ojanen, V., 2022. Sustainability-oriented innovations in smart cities: A systematic review and emerging themes. *Cities*, 126, 103716.

UN General Assembly. *Transforming our world: The 2030 Agenda for Sustainable Development*, A/RES/70/1, 21 October 2015, United Nations, New York, https://www.refworld.org/legal/resolution/unga/2015/en/111816 [accessed 20 November 2024]

Valencia, S. C., *et al.*, 2019. Adapting the sustainable development goals and the new urban agenda to the city level: Initial reflections from a comparative research project. *International Journal of Urban Sustainable Development*, 11 (1), 4–23.

van der Pluijm, R., and Melissen, J., 2007. *City diplomacy: The expanding role of cities in international politics*. The Hague: Netherlands Institute of International Relations "Clingendael".

van Twist, A., Ruijer, E., and Meijer, A., 2023. Smart cities & citizen discontent: A systematic review of the literature. *Government Information Quarterly*, 40 (2), 101799.

Vargas-Hernández, J. G., Pallagst, K., and Zdunek-Wielgołaska, J., 2023. Urban green spaces as a component of an ecosystem. *In:* Dhiman, S., ed., *Sustainable development and environmental stewardship*. Cham: Springer International Publishing, 165–198.

Vasist, P. N., and Krishnan, S., 2023. Fake news and sustainability-focused innovations: A review of the literature and an agenda for future research. *Journal of Cleaner Production*, 388, 135933.

Venetoulis, J., and Talberth, J., 2008. Refining the ecological footprint. *Environment, Development and Sustainability*, 10 (4), 441–469.

Volpi, V., Palatucci, M., and Marinelli de Marco, G., 2015. The emergent city. Interactive relational systems between public administration and citizen to foster sustainable processes of urban development. *Acta Europeana Systemica*, 5, 11–18.

Wakely, P., 2020. Partnership: a strategic paradigm for the production & management of affordable housing & sustainable urban development. *International Journal of Urban Sustainable Development*, 12 (1), 119–125.

Watkins, J., 2015. Spatial imaginaries research in geography: Synergies, tensions, and new directions. *Geography Compass*, 9 (9), 508–522.

Watson, V., 2016. Locating planning in the new urban agenda of the urban sustainable development goal. *Planning Theory*, 15 (4), 435–448.

Westman, L. K., Castán Broto, V., and Huang, P., 2019. Revisiting multi-level governance theory: Politics and innovation in the urban climate transition in Rizhao, China. *Political Geography*, 70, 14–23.

Xue, Y., *et al.*, 2020. Multi-sector partnerships in the urban development context: A scoping review. *Journal of Cleaner Production*, 268, 122291.

Yiwei, C., and He, W., 2020. The Education of human capital and the development of urban economic resilience. *In:* Yinhai Qian ed., *2020 4th International seminar on education, management and social sciences (ISEMSS 2020)*, 17.07.2020–19.07.2020. Paris, France: Atlantis Press.

4 Global, International Legal Frameworks Related to Sustainable Development

4.1 International Instruments: 2030 Agenda, the New Urban Agenda, and the Paris Agreement

As introduced in the previous chapters (Chapters 1 and 2), on the international level, three key instruments adopted in 2015 and 2016 have shaped the context for urban sustainable development: the 2030 Agenda with the Sustainable Development Goals (SDGs), the New Urban Agenda, and the Paris Agreement. They have almost global coverage and address policy- and decision-making at several governance levels. In this chapter, we argue that their legal character and respective approaches—with the 2030 Agenda and the New Urban Agenda on one side, and the Paris Agreement on the other side—differ substantially and explore the reception of the international instruments and their impact on the discourses on the local level (Section 4.1). While Chapter 2 explores the three instruments' approaches to the contexts of sustainable development, in this chapter, we provide an overview on their legal implications and key contents on participation that are relevant to understanding the backdrop for national legislation and policies (Section 4.2) as well as the local specifications of implementation (Section 4.3).[1]

4.1.1 The 2030 Agenda for Sustainable Development

The 2030 Agenda was adopted by the UN General Assembly (UNGA resolution 70/299) in September 2015 along with a set total of 17 SDGs (2030 Agenda, 2015). As a UNGA resolution, the text is not legally binding but it is widely accepted, which implies a strong political commitment by the supporting states. For the context of our research on sustainable urban development, the goals for sustainable cities and communities (SDG 11), and climate action (SDG 13) are of particular relevance. Other SDGs also address important aspects in the urban context, such as clean water and sanitation (SDG 6); affordable and clean energy (SDG 7); industry, innovation, and infrastructure (SDG 9); and responsible consumption and production (SDG 12). Building on the various SDGs, states are creating and implementing new targets and commitments, also linked to their respective obligations under other international and multilateral agreements on SDG-related topics.

DOI: 10.4324/9781003284864-4

The implementation takes place on the national level, and—while it is not centrally monitored and has no compliance mechanism—information about the progress is gathered on the international level: The UN Secretariat provides reports to the UN High-level Political Forum on Sustainable Development (HLPF) which meets annually under the UN Economic and Social Council (ECOSOC) (UNGA resolution 67/290). The HLPF replaces a previous Commission on Sustainable Development and was first convened in 2013. Every four years, the forum gathers heads of state and government, and the second SDG Summit took place in September 2023. The forum also provides regular reviews of the SDGs, building on input from the signatory states.

A key information tool is the "voluntary national reviews" (VNRs). The VNRs are provided by the respective government's views, so they have the character of national reporting and are not subject to external peer-review, but they can also include perspectives of relevant associations and civil society input. Four of the five states of our case study cities have provided VNRs twice since 2015. As the overview that follows will show, regarding monitoring, participation, and inclusiveness, they flag the difficulties and potentials for improvement, including for local indicators and participation opportunities (Finland), adopting integrated policies and plans towards inclusion (Iceland), a lack of new implementation tools for the Building and Planning act (Norway) and a shortage of affordable housing (Iceland, Norway, Sweden).

Finland's VNR from 2020 states that cities "work extensively with sustainable development, but not all cities link their work explicitly with the 2030 agenda" (Section 5.4.1). It points out *inter alia* that the SDGs do not form the starting point for the strategy or plans for most municipalities but are considered a useful framework for communication and outreach (Section 5.4.3). The VNR also notes "room for improvement" as the design of national indicators often does not allow them to be used for local monitoring, thereby adding the need to create indicators and monitoring on the local level (VLR). Some cities started working on "voluntary local reviews". On SDG 11, the VNR 2020 includes that participation opportunities for local residents have been increasing, but there would be additional room for improvement.

In Iceland's VNR from 2023, the SDG implementation on almost all SDG 11 indicators is assessed to be "close to achieving target", including 11.3 that sustainable development is the overriding concern of public policy in planning and construction, which has been included *inter alia* in the objectives of the Planning Act and as an aim of the national planning strategy and land use strategies at the local level. Two indicators are only showing "some progress towards target". One of these (11.b) is about adopting and implementing integrated policies and plans towards inclusion, resource efficiency, mitigation and adaptation to climate change, and resilience to disasters. In an Icelandic context, this includes risk assessments of avalanches, volcanic eruptions and fresh water and coastal floods. The other (11.6) is about the reducing the adverse per capita environmental impact of cities, with a special attention to air quality and municipal and other waste management. Acts and strategies working towards that are the National Air Quality Plan (2018–2029) and the Waste Management Strategy (2021–2032). Civil society's assessment is particularly concerned with housing shortages and transport infrastructure.

Norway's VNR from 2021 in contrast shows some challenges in the context of indicator 11.3 and points out that, for instance, the Planning and Building Act "needs to be supplemented with new implementing tools", to be fit for purpose (p. 76). Additional challenges are similar to Iceland in the context of disaster prevention (11.5) and local air pollution (11.6). The impacts of flooding, landslides, and heatwaves have been identified as a main challenge, as well as affordable and accessible housing for low-income groups, the elderly, and people with disabilities; in remote areas, population decline, together with a rising elderly population, has been identified.

Sweden's VNR from 2021 highlights for indicator 11.3 its policy instruments and coordinating fora to implement sustainable development, including the "Strategy for Liveable Cities" and its implementation of the "Policy for Designed Living Environment" (see also Section 4.2.5). A "Council for Sustainable Cities" provides coordination capacity for the work on sustainable development. Challenges in implementation are similar to the other Nordic countries: a shortage of housing, and also the management of existing infrastructure including roads, housing, and cities (p. 105).

Greenland did not submit a VNR, and the Danish VNR does not cover Greenland.

This overview of the VNRs illustrates the broader shared challenges in the countries of our case study cities, which are specified in their respective context in the following subsections 4.2 and 4.3. Given the responses from our interviewees in the case studies, local discourses focused more on urban planning, environmental impact assessments for infrastructure development and some specific sectors such as energy and waste management; they did not revolve as much around disaster risks and air quality.

4.1.2 The New Urban Agenda

The New Urban Agenda was adopted at the UN Conference Habitat III in October 2016 and endorsed by the UN General Assembly in December 2016 (Resolution A/RES/71/256, an edited volume is available by UN Conference on Housing and Sustainable Urban Development 2016). In addition to UN member states, international organisations and agencies, as well as subnational and local governments and their networks, participated in the preparation of the document, so it reflects different levels of input by actors in local governance, particularly in contrast to the Paris Agreement (see Section 4.1.3). While the resolution is—parallel to the 2030 Agenda—not legally binding, it does give internationally agreed upon guidance on what cities are expected to deliver in the context of urban sustainability. In line with the 2030 Agenda SDG11 (on inclusive and resilient cities), the New Urban Agenda addresses what defines liveable cities. Several parts of the document address planning and participation in different contexts, which are outlined below to provide the context for the next sections. It must be noted that many commitments overlap and are reiterative in their approach to be fully inclusive in several contexts. The full content of the New Urban Agenda with its additional aspects, e.g., of social coherence, disaster risk reduction, and investment, goes beyond the scope of this introduction.

The New Urban Agenda itself is structured in two main parts, the "Quito Declaration on Sustainable Cities and Human Settlements for All" (paras. 1–22) and the "Quito Implementation Plan for the New Urban Agenda" (paras. 23–175). The first part of the New Urban Agenda—the Quito Declaration—sets out a shared vision with eight elements that cities and human settlements should fulfil (para. 13 (a)(h)): in short, this includes the cities' social functions; being participatory, inclusive and accessible; achieving gender equality; meeting current and future challenges and opportunities; fulfilling territorial functions; promoting age- and gender-responsive planning and investment; adopting and implementing disaster risk reduction and management; and protecting, conserving, restoring, and promoting their ecosystems. This vision is underpinned with principles (para. 14) with (a) social, (b) economical, and (c) environmental dimensions. The first principle focuses on leaving no one behind and includes in our specific context "ensuring public participation, providing safe and equal access for all" (para. 14 (a)).

The Quito Declaration also adds three broad commitments by the signatories (para. 15) to implement the vision and the principles. The commitments relate to

- Becoming active at all stages of cities' work: planning, financing, development, governance, and management;
- Recognising the roles of all key stakeholders: national, subnational, and local governments, as well as civil society and other relevant stakeholders; and
- Adopting sustainable and integrated approaches to urban and territorial development when implementing their policies, strategies, capacity development, and actions on all levels.

Specifically, the second and third commitments underline the importance and broad understanding of (urban) stakeholders and inclusiveness in both the definition and implementation of urban policies and put the focus of all approaches to also be "people-centred, age- and gender-responsive" (para. 15 (b) and (c)). The call for action (paras. 16–22) also includes the affirmation to implement the New Urban Agenda in a participatory and people-centred manner (para. 16).

The second part of the New Urban Agenda—the Quito Implementation Plan—is itself structured in three sections: transformative commitments for sustainable urban development (paras. 24–80), effective implementation (paras. 81–160), and follow-up and review (paras. 161–175). With our focus on participation and involvement in planning, the elements of the previously described Quito Declaration are picked up in more specified commitments, which are briefly summarised for an overview below.

- In the first section, this is addressed in the context of "sustainable urban development for social inclusion and ending poverty", such as housing policies (paras. 31, 33); safe and inclusive, accessible, green, and quality public spaces (para. 37);

natural and cultural heritage (para. 38); a generally safe, healthy, inclusive, and secure environment (para. 39); and institutional, political, legal, and financial mechanisms (para. 41). For the context of "sustainable and inclusive urban prosperity and opportunities for all", the Quito Implementation Plan encourages overall effective participation and collaboration of all relevant stakeholder (para. 48) and highlights the participation of youth in particular (para. 61). In the context of "environmentally sustainable and resilient urban development", it underlines the participation in urban and territorial planning processes (para. 72).

- The second section on implementation contains the most references to participation and planning in the Quito Implementation Plan. This includes aspects of participatory planning and management of urban spatial development and urban policies (paras. 81 and 86), also highlighting the needed strengthening of local capacities to implement effective governance (para. 90), and partnerships between governments and civil society (para. 92). Further aspects cover the inclusiveness in planned urban extensions and infill (para. 97), adequate housing (para. 105), access to mobility and transport systems (para. 114), sustainable and climate resilient infrastructure and service provision systems (para. 119), cultural heritage, including indigenous peoples and local communities (para. 125), transparent and accountable expenditure control and sustainable financing (paras. 138–140), capacity development (paras. 148 and 155), information and communications technology policies and e-government strategies, including digital governance tools and data platforms (paras. 156 and 160).
- In the third section, signatories are encouraged to a voluntary, country-led, open, inclusive, multilevel, participatory, and transparent follow-up and review of the New Urban Agenda, which also mentions the involvement of and partnerships with all relevant stakeholders (para. 162).

4.1.3 The Paris Agreement

The Paris Agreement (Paris Agreement 2015) is an international treaty with 195 Parties (June 2024). As such, it differs substantially from the 2030 Agenda and the New Urban Agenda in both its approach on national commitments and the participation of cities and local actors. The Paris Agreement was negotiated until 2015 under a work stream of the United National Framework Convention on Climate Change (UNFCCC 1992) and is building on the UNFCCC's institutions and processes. At the same time, it is adding new obligations (e.g., Nationally Determined Contributions, NDCs) and topics that were either non-existent or not as prominent in the UNFCCC. The Paris Agreement entered into force in 2016 when the necessary thresholds of contracting parties and global greenhouse gas emissions covered by them was surpassed.

At the agreement's core is an ambition mechanism that requires the contracting parties (states and the European Union) to regularly provide updates regarding their national ambitions and to report about their implementation of respective policies. Key mechanisms of the UNFCCC (including its financial mechanisms) continue to exist and often serve both—the UNFCCC and the Paris Agreement. Under the

Paris Agreement, the parties have established a new detailed system of transparency reporting (enhanced transparency framework) in 2018 that requires, in principle, that all parties report in "biannual transparency reports" from 2024 on. The reporting will be subject to a country-specific expert review. In addition, the Global Stocktake (GST) under the Paris Agreement takes place every five years, providing information on the collective progress of all parties. The result of the first GST in 2023 included a focus on keeping the 1.5°C temperature increase within reach by setting more ambitious political goals for all parties, including the "transition away from fossil fuels" ("UAE Consensus", Decision 1/CMA.5). However, the GST also does not single out individual parties, which is in line with the focus of the Paris Agreement on party-driven and bottom-up implementation. Lastly, the Paris Agreement's Compliance Committee (PAICC) can only engage in specific cases, for example, states not providing a new NDC, failing to submit specific reporting or having significant and persistent inconsistencies in their transparency reporting. The PAICC is taking a facilitative and non-punitive approach. It does not monitor the level of parties' ambition and has its focus on procedures that help parties fulfil these obligations.

As a treaty under public international law, the Paris Agreement addresses states. Cities are only included as a part of the subnational structures and not mentioned at all in the Agreement. The Paris Decision, which is adding details to the Agreement's interpretation, but was still adopted under the UNFCCC, mentions cities. However, here they are included only in the context of welcoming the efforts of "non-party stakeholders" to address and respond to climate change (Decision 1/CP.21, paragraph 133). The broader term "subnational level" is used in the Agreement in the contexts of adaptation (Article 7.2) and capacity building (Article 11.2), but in contrast not used for aspects such as public participation (Article 12).

Regarding the involvement of cities, their standing in the process is at the same level as other constituencies, such as environmental NGOs, trade unions, business NGOs, and youth NGOs. The Agreement recognises the role of observers from governmental and non-governmental organisations, which can be present in negotiations but have only a rather limited procedural role to listen in and—in a few cases—voice their priorities; they are not involved in the decision-making. Among the recognised observer constituencies are also the local governments and municipal authorities (LGMA), which is one of the few indications of their key role in implementing the contracting parties' obligations.

4.2 Reception of International Instruments and Discourses on Sustainability in the Case Study Cities

The international instruments outlined above are generally translated through national legislation and policies to the municipality level. The perception of the international instruments outlined above, however, varies substantially on the local level. Our interviewees provided different aspects of relevance for the case study cities.

Firstly, the interview feedback indicated that the concept of sustainable development is present in policy- and decision-making and to a varying degree in the discussions taking place at the local level. In Sweden, for instance, the discussion on sustainable development was perceived as strengthening the work in Kiruna and also as present in Luleå (researcher, Kiruna/Luleå, 20.01.2022 and business representative, Kiruna, 16.03.2022). More critically, it was noted that "no one really knows what sustainability means in certain contexts" (researcher, Kolari, 13.01.2022). The importance of national support for implementing the international instruments is recognised, for example, through research projects at universities (researcher, Tromsø, 08.03.2023; researcher, Kolari, 13.01.2022; researcher, Rovaniemi, 20.01.2023). In contrast, however, an interviewee in Iceland responded that while the importance of sustainable development would be recognised, the corresponding requirements would not often be followed, due to economic reasons or beliefs that space and resources are limitless (researcher, Iceland, 15.09.2022). Also, the impact of the COVID-19 pandemic in 2020 and the following years led in some cases to a different focus in planning (business representative, Akureyri, 28.04.2023).

This observation links to the second aspect: local priorities are likely to override the international discussion in case of a direct conflict, for example, in the case of Kolari where the discourse did not change much after the signing of the Paris Agreement (business representative, Rovaniemi, 18.02.2022) and the debate between natural values and business development has been heated (former elected official and business representative, Kolari, 04.03.2022). In some cases, this aspect included the perception that joining environmental agreements in particular, for example, being bound by rules of the International Whaling Commission, could negatively impact welfare or economic development (researcher and representative of the administration, Nuuk, 16.08.2023). One suggestion was to keep the benefits (e.g., tax revenues) within the community to also gather local support, for example, for green energy projects (politician, Rovaniemi, 24.11.2022).

Thirdly, the remoteness of the (smaller) Arctic communities can lead to a sense of not being able to make a difference and also to be "at the receiving end of global developments" (researcher and resident, Akureyri, 14.09.2022). This was mentioned for instance in discussions on whether Greenland should join the Paris Agreement, which they did in 2022 (researcher and representative of the administration, Nuuk, 16.08.2023) but also in the general context of "people in the periphery" (researcher and resident, Akureyri, 14.09.2022).

Fourthly, however, our interviewees highlighted the international recognition of sustainable development as contributing to a positive city "brand" and potentially leveraging a city's visibility, which could specifically benefit smaller and more remote cities. In this way, local ambition is linked to international expectations (researcher, Akureyri, 27.05.2022) and also used in international exchanges, for example, via the Arctic Mayors Forum and in Tromsø's (unsuccessful) application to join the 100 climate neutral cities initiative (politician and researcher, Tromsø, 10.03.2023).

We provide further insights on how the pursuit of sustainable urban development at the local level is discussed in the cities under analysis here in Section 4.3, but the following section discusses first how the goals from the global instruments are translated at the national level via constitutions, legislative acts, and policies.

4.3 National Translation of Urban Sustainable Development: Constitutions, Legislative Acts and Policies

The first step needed to translate the international instruments to the local level takes place on the national level. The respective constitutional provisions, as well as acts on the organisational arrangements and on the substance (planning, environment, impact assessments), provide the legal framework in which the municipalities operate on the local level. The web of legal provisions applicable in the case study cities that can potentially impact the sustainability of urban planning is complex and cannot be covered in its entirety. We attempt to provide in this section an overview that links the national level with the regional level, before discussing the local discourse on urban sustainable development below (in Section 4.3). The administrative application and implementation practice of these governance arrangements are then elaborated in further detail in Chapter 5.

Table 4.1 provides an initial overview on the key constitutional provisions and implementing acts for all five countries.[1]

4.3.1 Finland

In Finland, the Constitution (Suomen perustuslaki, Act of 11.6.1999/731) sets out the basic structure for municipal and other regional self-government which is then elaborated on in several acts. The constitutional provisions most relevant for sustainable urban development emphasise the importance of environmental protection (Section 20) and allow for local self-governance (Section 121).

- Section 20 sees it in the responsibility of public authorities to guarantee everyone the right to a healthy environment and to give everyone the possibility to influence the decisions that concern their own living environment.
- Section 121 provides powers to the municipalities, in particular granting them the right to levy taxes and fees to finance their activities. They have their tasks assigned per law. A special reservation for the Sámi reserves their autonomy regarding their language and culture in their homeland, in accordance with the law.

As an EU member state, Finnish national law implements, to a large extent, EU environmental directives and regulations. National legislation linked to the implementation of the two constitutional provisions above includes in particular:

- The Local Governance Act (10.4.2015/410, Kuntalaki, repealing the previous Act 365/1995), sets the key legal framework for the municipal organisation, ranging from elected bodies and personnel, to cooperation between municipalities and municipal finances. The Act also includes provisions that provide context for the planning and participation of their citizens. In general, the municipalities have the duties to carry out their tasks under self-government and those assigned by law (Section 7). The organisation of the services needs to provide inter alia equal access (Section 8 para. 2 (1)). The general right of

Table 4.1 Country overview on key constitutional and legal provisions on the national level with relevance to sustainable urban development, in particular local planning and participation

Country	*Constitutional provision*	*Main content*	*Implementing legislation*	*Relevant sections*
Greenland (part of the Kingdom of Denmark)	–	Local self-government	Greenland Self-Government Act (Act No. 473 of 12 June 2009) (Selvstyreloven)	Part Four, Chapter 1, Section 3 and Part Four, Chapter 2
		Planning and building	Act on land use and urban planning and housing (Inatsisartutlov om planlægning og arealanvendelse)	Sections 1, 5, 7, 11, 14ff., 21f., 23, 34ff., 36ff., 56f.
		Environmental protection	Environmental Protection Act (Inatsisartutlov nr. 9 af 22. November 2011 om beskyttelse af miljøet)	Sections 2, 4, 8, 9, 41, 42, 44, 45
Finland (Suomen perustuslaki)	Article 20	Right to a healthy environment	Environmental Protection Act (No. 527/2014)	Chapter 4, Section 34; Chapter 10a, Section 115a; Chapter 11, Section 116; Chapter 15, Section 143; and Chapter 20, Section 202
			Act on the administration of environmental protection in municipalities	Section 6
			Land Use and Building Act (No. 132/1999)	Chapter 1, Section 5; Chapter 2, Section 20; Chapter 5; and Chapter 7
			Nature Conservation Act (No. 9/2023)	Section 6
			Act on Environmental Impact Assessment (No. 200/2005)	Sections 2, 4, 5, and 6
	Article 121	Municipal self-government	Local Government Act (Act No. 410/2015) (Laki kunnallisesta itsehallinnosta)	Sections 8, 22, 26, 27, 28, 37, 120a, 134ff.

Iceland (Stjórnarskrá Íslands)	Article 78	Municipal self-government	Local Governance Act (no. 138/2011)	Chapters I, IX (Artt. 102 and 103)
			Organizational Law (no. 123/2010)	Artt. 1, 4, 6, 10, 12, 21, 28, 30, 37, 40, 45
			Act on policies and action plans in the field of housing and planning, transport and regional affairs (no. 30/2023)	Articles 3 and 5
Norway (Grundloven)	Art. 108	Recognition of Indigenous People	The Planning and Building Law (Lov om planlegging og byggesaksbehandling	Chapters 3-1 and 5-4
			Nature Diversity Act (Lov om forvaltning av naturens mangfold—naturmangfoldloven	Sections 14 and 43, Para. 2
	Article 112	Rights to a healthy environment and natural environment	Planning and Building Act (Plan- og bygningsloven)	Chapters 10–13 (Sections 10-1 and 11-2)
			Nature diversity Act (Lov om forvaltning av naturens mangfold—naturmangfoldloven)	Sections 34, 41, 43, 62
Sweden (Regeringsformen)	Chapter 1, Sections 2	Sustainable development, good environment for present and future generations	Planning and Building Act (Plan- och bygglag)	Chapter 1, Sections 1 and 2; Chapter. 3, Sections 8 and 12; and Chapters 4 and 5, Section 10a–f
			Environmental code	Chapters 3–6

Source: Authors' compilation.

participation is included in the sections on voting and additional opportunities for participation and influence (Sections 20ff.), referenda (Sections 24f.) and the involvement in various councils, including youth, elderly and disability councils (Sections 26 to 28). The council has to develop a municipal strategy (Section 37), which should include opportunities for municipal residents to participate and influence. An appeal is possible against decisions by the municipal board or a committee (Sections 134ff.).

- Land Use and Building Act (5.2.1999/132, Maankäyttö- ja rakennuslaki): This act regulates land use planning and construction in Finland. It sets out the responsibilities of municipalities and regional councils in planning and managing land use and establishes procedures for the preparation and approval of land use plans on several levels (including provincial and town levels). Among its objectives of spatial planning (Chapter 1, Section 5) are aspects of a needs-oriented environment, environmental protection, and the sparing use of natural resources, among others. Municipalities are responsible for the planning of the use of areas and the implementation of land policy in their respective areas (Chapter 2, Section 20) and also must have "sufficient resources and expertise" at their disposal: at over 6,000 inhabitants, a municipality is required to have a "zoning officer". The planning usually involves two levels, a more general or master plan (Chapter 5) and more detailed site plans (Chapter 7). A site plan that has not yet been implemented to a large part may require an assessment of whether it is still up to date if it has been in force for more than 13 years and has not already been assessed in the last five years—however, there is no appeal possible against the municipality's decision (Chapter 7, Section 60). The act will be replaced by Act 752/2023—the Area Use Act from 1 January 2025 on.
- Environmental Protection Act (27.6.2014/527, Ympäristönsuojelulaki): This act sets out measures to protect the environment, including provisions on environmental permits (Chapter 4) and procedures (Chapters 5–9), as well as on the substance of different aspects of nature conservation, for example, on the state of the environment (Chapter 15). The act also outlines the responsibilities of public authorities in implementing the constitutional right to a healthy environment: municipalities can decide—subsidiary to national authorities—in certain cases on permit applications, for example, for activities that do not have significant environmental effects and which do not affect an area beyond the municipality (Chapter 4, Section 34). The local authorities are also processing the notifications of risks of environmental pollution by facility operators (Chapter 10a, Section 115a, and Appendix 4) and registration notices by operators for specific activities (Chapter 11, Section 116 and Annex 2). Overall, municipalities are monitoring the state of the local environment in their area (Chapter 15, Section 143) and have to draw up local supervision plans (Chapter 18, Section 168). They are also foreseen to issue general regulations to implement the requirements of this act (Chapter 20, Section 202), with certain exceptions, for example, regarding activities that require a permit, notification, or registration. Municipalities need to inform about environmental protection regulations and make them publicly available, and—before issuing such regulations—need to

give the relevant national state supervisory authority and, if necessary, other authorities the opportunity to issue statements (Section 203).

- The Act on the administration of environmental protection in municipalities (24.1.1986/64, Laki kuntien ympäristönsuojelun hallinnosta) summarises the tasks of the designated municipal authority, which include among others the implementation of prescribed tasks, the planning and development of environmental protection and monitoring of the state of the environment (Section 6, see also above regarding the Environmental Protection Act).
- In addition, the provisions of the cross-cutting Nature Conservation Act (a new act 5.1.2023/9 entered into force on 1 June 2023: 5.1.2023/9, Luonnonsuojelulaki) and the Act on Environmental Impact Assessment of Plans and Programmes of the Authorities (8.4.2005/200, Laki viranomaisten suunnitelmien ja ohjelmien ympäristövaikutusten arvioinnista) as well as sectoral legislation such as the Water Act (27.5.2011/587, Vesilaki) and the Waste Act (17.6.2011/646, Jätelaki) can impact planning procedures. Notable for the new Nature Conservation Act is a new provision on the protection of Sámi culture (Section 6), which includes the promotion of the conditions for practicing Sámi culture, where possible. Also, under the Environmental Impact Assessment Act, the authority that is responsible for a plan is also responsible for the environmental impact assessment (Section 6), which adds to the tasks of the local municipalities.

In addition, the national level launched programmes to support exchange among cities, including on the international level, as well as to develop tools to improve their implementation of sustainability targets:

- The Sustainable City Programme ran from 2019 to 2023 and lists 89 Finnish participating cities and municipalities (Kestävä kaupunki 2023a). Its focus is on exchange of information and the development of tools to support cities in the implementation of the New Urban Agenda and the 2030 Agenda. Thematic areas include: (1) carbon reduction (e.g., by city planning, transport arrangements, public procurements); (2) smart cities (e.g., by digitalisation and mobility services); (3) social sustainability (via access to education and healthcare); and (4) health (e.g., on air quality, noise exposure and green spaces) (Kestävä kaupunki 2023b). From the two Finnish case studies, only Rovaniemi participated in a project under this programme (see Section 4.3.1).
- The MayorsIndicators tool (MayorsIndicators 2023) was developed to monitor and compare the SDG indicator implementation between Finnish, Swedish, and UK municipalities. It assigns scores to the relative performance of the respective municipalities from 100 to 0 per indicator. The outcomes are added for each municipality and the outcomes grouped into five categories. While the highest-ranking municipalities score over 1,000 points (e.g., Luleå in and Kiruna in Sweden, see below), Rovaniemi ranks in the midfield (third category) with 853.67 points while Kolari is only at the lower end of the scale (fourth category) with 794.83 points. More detailed results are available to the members of the network.

From the constitution and national level legislation, Finnish municipalities have wide-reaching competences and obligations for local planning, including several layers of planning (master plans, local plans), for environmental impact assessments, and for approving certain permits. The national level supports the local level with programmes and tools for knowledge exchange and monitoring in the context of (urban) sustainability.

The Regional Council of Lapland reviewed the municipal strategies of Lapland's 21 municipalities in 2019 with regard to the implementation of the 2030 Agenda for Sustainable Development and the Lapland region's vision for sustainable regional development (Finland VNR 2020). It found that sustainability is "strongly embedded in the strategies, although the objectives identified to promote sustainable development may differ considerably" (Finland VNR 2020). From our interviewees in the Finnish case study cities (Rovaniemi and Kolari), we received comments that the local planning is envisioned as being very participatory and contains plans for every kind of process, including strategy papers, but the implementation of these plans seems to be falling short (researcher, Rovaniemi, 28.01.2022). In the context of participation, interviewees noted that while procedures were followed, the involvement could be formalistic, with no efforts going beyond the minimum requirements, and also require expertise on the side of citizens to understand when the right moment to participate would be (researcher, Rovaniemi, 28.01.2022). One comment also noted that appeal mechanisms could be used to slow planning processes down (elected official, Rovaniemi, 24.11.2022). A more detailed analysis is presented in Chapter 5.

4.3.2 Greenland

The legal framework in Greenland is different from the other countries of our case studies in that it is part of the Kingdom of Denmark, but makes and implements its own legal system under the Self-Government Act (Lov om Grønlands Selvstyre), which builds on the 1979 Home Rule Agreement and expands the Greenlandic competences. While discussions on a Greenlandic constitution have been held for decades, and a first draft was published in April 2023, for the time being, the legal order is still unchanged. Greenland is not part of the European Union—other than Denmark—so also the EU legal order does not apply directly.

The Self-Government Act is based on an agreement between Naalakkersuisut (Government of Greenland) and the Danish Government, and was approved by a referendum in Greenland in November 2008 and adopted by the Danish Parliament (Folketing) in June 2009. Key policy areas, including foreign policy and security, as well as law enforcement, are still under responsibility from Denmark. Many competences for the local level, however, have been transferred to the Greenlandic parliament (Inatsisartut) and government so that the Danish laws on these issues are not further elaborated here. The Self-Government Act differentiates between five fields of responsibility that are transferred at points of time fixed by the Self-Government authorities (list 1 of the schedule) and 28 fields that require prior negotiation with the central authorities of the Realm (i.e., Kingdom of Denmark) (list 2 of the schedule). List 1 of the schedule in the annex to the act includes within the context of urban planning, for instance, the road traffic area and the law of property and obligations; list 2 includes a much longer list, including for example the administration of justice, financial regulation and supervision, and the mineral resource area.

Some key legislative acts on the national level, relating to aspects of sustainable development, planning, and participation, include:

- The Act on land use and urban planning and housing from 2010 (Inatsisartutlov om planlægning og arealanvendelse) has been updated most recently in November 2023. It stresses among its aims for nature protection, socially appropriate distribution between open areas and built areas, promotion of business, social and environmentally sustainable development, and public participation in the land use planning (Section 1). The national government provides a digital, geographic information system with a planning portal (Section 5) and ensures coordination of national planning with municipal planning and sector planning (Section 7). In this context, the national authorities have the opportunity to object to municipal plans that conflict with special considerations on the national level (Section 11). Overall, however, the municipal planning lies in the hands of the respective municipal council (Section 14ff.), including also perspective plans for sub-areas for future desired use (Section 14a). Local plans can elaborate more detailed rules (Sections 21f.). Plan proposals must be published and made available to the public and presented for review for at least 8 weeks (Section 23, Paras. 1 and 3). Also, the municipal council needs to conduct an information campaign, aiming to enable a "public debate about the objectives and content" of the plan proposal (Section 23, Para. 5). For the broader context, the municipal board must publish within its first half of its election period a proposal for its strategy regarding municipal planning (Section 34ff.). It is noteworthy that it is not possible to buy and sell land in Greenland, but only to get rights to use a specific plot of land, a so called "area allocation", which can allow one to build, change, or demolish permanent buildings or facilities or change their use (Section 36ff). Regarding land allocations, the act clarifies that no area may be excluded from general use and put into use without permission from the land use authority, which is the municipal board for its respective municipality (Sections 37 and 38, Para. 1). There is an exception for permissions by the national government, e.g., for areas outside the municipal division, for facilities of national planning importance, and land use in accordance with national building laws and planning directives (Section 38, Paras. 2–4). Individuals have the right to appeal decisions to the national government within 8 weeks after the municipal decision has been announced (Sections 56f.).
- The Environmental Protection Act (Inatsisartutlov nr. 9 af 22. November 2011 om beskyttelse af miljøet) takes a wide approach and includes regulations on the pollution of air, water, ice, mountains, and soil, as well as on noise pollution, limiting waste of resources and on the promotion of recycling (Section 2). It does not, however, cover environmental protection in connection with

legislation on natural resources (Section 4). The central authority is the government of Greenland, the local environmental authority is with the respective local council (Sections 8 and 9). The act also contains requirements for an environmental impact assessment of major building and construction works that can be assumed to have a significant impact on the environment (Sections 41 and 42). Additional Government orders lay down which installations and facilities are covered, and the specific requirements, for instance in Order No. 5 on the assessment of impact in construction on the environment and payment for environmental surveillance (Selvstyrets bekendtgørelse nr. 5 af 27. marts 2013 om vurdering af visse anlægs virkninger på miljøet og betaling for miljøtilsyn). Overarching pollution plans are in the general responsibility of the national government (Section 44), but it can order e.g., municipal councils to draw up action plans for specified business activities (Section 44 para. 2). The national government also prepares an overall waste plan (Section 45).

The Greenlandic case is special in comparison for several reasons: Greenland does not have its own constitution, the competences for different policy fields are distributed between Greenland and Denmark and—as the only case study city in our research—the city of Nuuk is not only an administrative centre of the municipality Sermersooq but also the national capital. In the legal context of urban sustainable development and participation, however, these differences are less relevant. The municipality has—similarly as in the other countries of our case study cities—the competence and duty of local planning, permissions for land use and impact assessments. Overall, the national level provides supporting tools to municipalities, for example, a geographic information system and planning portal. Within the city of Nuuk, it is important to keep in mind the different authorities working on the municipal level and for the national government, which has several overriding competences, for example, for facilities of national planning importance.

On the legislative context, one of our interviewees remarked that it can be difficult "to enforce and control the development with a view to sustainability", as the municipal level might not always have the tools to stop non-sustainable developments (administration, Nuuk, 31.03.2022). This requires a closer look into the local contexts of implementation (see Section 4.4.2 and Chapter 5).

4.3.3 Iceland

In Iceland, the link between the national level and municipalities is the main focus because Iceland does not have regional governments. Icelandic municipalities exercise to a large extent self-autonomy within legal frameworks described below. Key services for their inhabitants in the context of urban sustainable development include clean water, heating, sewage, and other basic infrastructure, as well as waste management, planning matters, and building inspection (see Iceland's VNR 2021, p. 49). The Association of Local Authorities represents them on the national level.

The Icelandic constitution links to municipalities only in its Article 78. It provides that "local authorities shall govern their own affairs themselves as provided by law. The revenue sources of local authorities shall be determined by law, as shall their right to decide whether, and to what extent, to exploit them". The national laws are decided in the Parliament (Alþingi, 2023). Three key implementing provisions on the national level elaborated here are the Local Government Act (no. 138/2011, LGA), the Organizational Law (no. 123/2010), and the Act on policies and action plans in the field of housing and planning, transport and regional affairs (no. 30/2023). Additional laws with potential impacts on the municipal planning include the Law on Cultural Heritage (no. 80/2012), the Nature Conservation Act (no. 60/2013), and the Act on Hygiene and Pollution Prevention (no. 7/1998).

- The LGA outlines the link between the national level, meaning the respective ministry responsive for local government affairs and the municipalities. The ministry is required to submit at least every three years its proposals for a schedule of resolutions that relate to municipal affairs in the next 15 years and establish an action plan for the next five years. The LGA Chapter IX is on "collaboration between municipalities and agreements on the operation of tasks". Residents have the right to an opportunity to influence and participate in the municipality's administration and preparations of policy formulation (Art. 102). The LGA also adds an obligation for the respective municipal council to provide information to its residents on municipal affairs under consideration and determination. This includes information on plan on long- and short-term services of the municipality, its finances, its environment as well as on goals that have already been established (Art. 103).
- The Icelandic Organizational Law includes among its aims "to promote the rational and efficient use of land and land quality, ensure the protection of landscape, nature and cultural values and prevent environmental damage and overuse, with sustainable development as a guiding principle" (Art. 1 b.). The guiding function of sustainable development is repeated in the sections for the national planning policy (Art. 10 para. 2), organizational plans by municipalities (Art. 12 para. 5), and for the guidance by the National Planning Agency for the preparation of such organizational plans (Art. 45 para. 11). The act's definition of "sustainable development" reads as "Development that meets the needs of the present without reducing the possibilities of generations to meet their needs. This means that the pursuit of economic quality must go hand in hand with the protection of the environment and the basic quality of the earth" (Art. 2 no. 18). Municipalities are required to form a planning committee, to be elected by the local council (Art. 6) and have a planning obligation within their boundaries (Art. 12), usually consisting of a master plan setting a strategy for at least 12 years (Art. 28), and additional more detailed local plans (Art. 37). The proposals by the municipality, its assumptions and available environmental assessments must be presented to the residents and other interested parties (Art. 30 for master plans, Art. 40 for local plans). Regional plans are possible to enable coordinated planning between two or more municipalities (Art. 21).

The National Planning Agency monitors the overall state of planning in the municipalities and has the task to assist and guide them in creating their organizational plans (Art. 4 c. and d.).

- In a rather recent development, the Act on Regional Planning and Municipal Planning (*Lög um byggðaáætlun og sóknaráætlanir* nr. 69/2015) was repealed in May 2023 by the Act on policies and action plans in the field of housing and planning, transport and regional affairs (*Lög um stefnur og aðgerðaáætlanir á sviði húsnæðis- og skipulagsmála, samgangna og byggðamála*, no. 30/2023). It allows the ministry on the national level to submit resolutions on transport, housing and regional policy for 15 years each with a corresponding action plan of five years (Article 3), to be complemented by three thematic councils (Article 5) which would then follow up with proposals for policies and action plans for the policy implementation. The Association of Icelandic Municipalities gets to nominate one member in each of these councils (Article 5 para. 3).

Additional governance elements in the context of implementing the legal provisions above include the national planning strategy and the (not yet finalised) urban policy:

- Iceland introduced its current national planning strategy in 2015, based on Articles 10 and 11 of the Planning Act (and further defined by the regulation on the national planning strategy). The strategy is valid until 2026 (12 years) and goes beyond the previous three-year strategy. Local authorities are to implement this strategy in their regional, municipal, and detail plans. It includes several aspects of sustainable development in subsections on vegetation, agriculture, energy sources, transport, resources, and also urban development (see Section 3.2). In this context, the strategy aims for an improvement of existing building areas, limiting them to protect surrounding nature and agricultural areas, and to reduce the societal costs of infrastructure, for example, through shorter distances and better public transport. Tourism development is linked to the acceptance of the local population and respecting the local circumstances.
- Iceland has not yet released a National Urban Policy but its planned development included in the regional development plan 2022–2036 (OECD 2023): this would aim to strengthen both Reykjavik as capital and Akureyri as a regional hub for services, culture, and job opportunities.

Also in Iceland, the municipalities are in control of the planning of their respective space, as it relates to their own affairs. While the Icelandic division of competences does not include a regional level, the organisation of municipalities into an association, which is also involved in several councils on the national level, provides them with additional leverage beyond the territory of a single municipality. Something that stands out in comparison with the other countries of our case studies are the long time periods required for the national planning relating to municipalities (15 years), as well as the national planning strategy (12 years), which should give the local level additional security of planning, and also the local level is obligated to set a strategy with their master plans for 12 years. Having said that, one of our interviewees highlighted that the rise of tourism in Iceland was very fast so that "the plan, the legislation was maybe not ready for it" (administration, Akureyri, 09.05.2023). Also here, the "how" of implementation in the local context is key (see Section 4.3.3 and Chapter 5).

4.3.4 Norway

The Constitution of the Kingdom of Norway (last consolidated in FOR-2023-05-26-739) includes the individual human right to a healthy environment and to a natural environment "whose productivity and diversity are maintained" (Article 112). It also gives citizens the right to information on the state of the natural environment to safeguard their right. For the Indigenous Sámi people, the state needs to create the conditions for them to preserve and develop their language, culture, and way of life (Article 108). Also in Norway, two acts are of particular importance to municipalities and their competence for sustainable urban development and planning.

- The Planning and Building Law (Lov om planlegging og byggesaksbehandling, LOV-2008-06-27-71, last changed by LOV-2023-06-16-73) provides for the municipal planning (Chapters 10–13), including its strategy, the master plan and zoning plans. Already on the level of the planning strategy, the municipalities should facilitate "broad participation and general debate" (Section 10-1). The overall master plan is then required to also include a "social component" (Section 11-2) which aims to address also "long-term challenges, goals and strategies", and shall be implemented with the participation of other public bodies and private individuals. The foundations for Sámi culture are mentioned later (Section 3-1) and the involvement of the Sámi Parliament is necessary if a planning proposal affects Sámi culture (see Section 5-4).
- The Nature Diversity Act (Lov om forvaltning av naturens mangfold—naturmangfoldloven) covers the possibilities to designate protected areas (Chapter V) which entails different types of protective levels. The king can delegate his power to a municipality as an authority under Chapter V (Section 62). Within the respective framework, the king may "by regulations prohibit or regulate activities or traffic that in themselves or together with other uses may counteract the purpose of the protection" (Section 34, Subsection 4). Regarding the process, the law requires inter alia that intended conservation proposals are made "with the best possible cooperation with landowners, licensees, affected business interests and representatives of the local population, including practitioners of Sámi culture and practitioners of Sámi industry, the municipality and county municipality, the Sámi Parliament and other relevant authorities" (Section 41, Subsection 1). It also requires to "clarify" the planning part with the municipality and county municipality if municipal and regional planning work begins simultaneously (Section 41, Subsection 2). The consultation also calls for publishing the proposal for inspection both in a physical location and in at

least one newspaper "that is commonly read on site" (Section 43, Para. 1). The involved municipality and county municipality get to comment, as well as the Sámi Parliament if the protected area affects Sámi interests (Section 43, Para. 2). This is in line with the overarching guidance that decisions directly affecting Sámi interests "due emphasis" is required to consider the natural basis for Sámi culture (Section 14).

Additional policies on the national level support the municipalities—both cities and rural areas—in the implementation of sustainable development in the local context.

- The white paper on "A good life throughout Norway—regional and rural policy for the future" is included in a ministerial report to the Norwegian parliament (Norwegian Ministry of Local Government and Regional Development 2023). It covers the years 2022–2023 and focuses on improving the living conditions and services in rural areas. As such it has some relevance for the case study city of Tromsø which is part of the wider Troms and Finnmark county (with 20 other municipalities). Generally, the white paper shows an initiative to bridge the divide between growing urban settlements and the rural areas. The government aims inter alia to "give municipalities room to manoeuvre" (Norwegian Ministry of Local Government and Regional Development 2023, p. 14) in the planning and cooperations with a number of initiatives, such as new guidelines for land use policy, developing rural growth agreements. It also addresses the accessibility of local services, the support for infrastructure and transport services, digital access via broadband and mobile coverage, business development and sustainability in e.g., food production and tourism, as well as educational and labour opportunities.
- An older policy, reflecting the developments in 2016–2017, on "Urban sustainability and rural strength—in brief", covered the connection between cities and the rural areas more specifically (Norwegian Ministry of Local Government and Regional Development 2017). As also specified later in the 2022–2023 white paper, already here, the white paper stressed the importance of equal living conditions, access to services, business development, infrastructure, and innovation. The previous 2016–2017 white paper still stressed in addition, however, more specifically the importance of planning based on the Planning and Building Act and aimed e.g., to enhance community planning through better guidance, enhance participation of citizens in the entire planning process and better coordination.

The Norwegian distribution of competences is parallel to the other case studies with municipalities having the competence and obligation to provide master plans and zoning plans. In addition, they have to involve the Sámi Parliament where a proposal could affect Sámi culture. The procedure is similarly laid out in the Nature Conservation Act for protected areas. From our interviewees, a distinction was drawn between the adoption of plans by the municipality and the preceding design of these plans, which could also be prepared by private developers. This would bear the risk of keeping citizens' participation out of the initial stage and include them only later via hearings and other procedural elements (researcher, Tromsø, 08.03.2023). In some sectors (e.g., tourism and industrial development), the participation would also be lacking. The implementation in the local context is described in more detail below (Section 4.4.4 and Chapter 5).

4.3.5 Sweden

The Swedish constitutional system grants special status to four laws, including the "instrument of government" (Regeringsformen, in: Swedish Code of Statues—Svensk författningssamling, SFS—1974:152, last changes by SFS 2022:1600), which includes inter alia the division of competences. Promoting sustainable development leading to a good environment for present and future generations is required for all public authorities (Chapter 1, Section 2).

Swedish local government includes both regions and municipalities. The municipal self-government is mentioned as a part of the implementation of the Swedish people's government (Chapter 1, Section 1) and the existence of municipalities at the local and regional level is guaranteed (Section 7). The basics of their organisation, their forms of operation and other powers and duties have to be issued by law (Chapter 8, Section 2). Key competences of municipalities are the management of local and regional affairs of public interest on the basis of local self-government, as specified by law, for which they also may levy taxes (Chapter 14). The municipalities' competence for local planning, infrastructure, and utilities puts them in the focus for our following overview.

As an EU member state, Swedish national law implements to a large extent EU environmental directives and regulations. Selected national legislation linked to the implementation of the constitutional provisions above includes in particular:

- The Planning and Building Act (Plan- och bygglag, SFS 2010:900, last changed by SFS 2023:173) puts the planning of the use of land and water in the responsibility of municipalities (Chapter 1, Section 2). The act has inter alia the purpose to promote "a good and long-term sustainable living environment for people in today's society and for future generations" (Chapter 1, Section 1). More specific requirements are set out by the Environmental Code (see below), which must also be followed by the municipality's masterplan (Chapter 3) and detailed development plans (Chapter 4) as well as their implementation. The development of a masterplan also requires a consultation including with members of the municipality, authorities, associations and other individuals (Chapter 3, Section 8) and a review period of the plan of at least two months (Chapter 3, Section 12). The more specific instrument of a detailed development plan is also subject to consultation and might be preceded by a planning statement (Chapter 5, Sections 10a-f) that also needs to be announced ahead of time. A detailed development plan is automatically "expected to have a significant environmental impact" if the planning is aiming for instance the use as an industrial area, shopping centre, but also more specific for tourist infrastructures such as ski slopes/lifts, or hotel complexes.

- The Environmental Code (SFS 1998: 808, last changed by SFS 2022:1799) specifies in the context of urban sustainable development the Planning and Building Act (see above). It contains rules on the management on land and water areas (Chapters 3 and 4), lists specific land-use interests, defines areas of national interest and sets out main environmental quality standards (Chapter 5). The rules on environmental impact assessments are elaborated in Chapter 6, covering both Strategic Environmental Assessments for plans and programmes, and Specific Environmental Assessments for activities and measures. The municipality is required to consult with other municipalities and authorities and make its decision publicly available.
- The Swedish "Policy for Designed Living Environment" is included in a national bill (2017/18:110) that replaced in 2018 the previous six national goals for the state's involvement in architecture, form ("formgivning") and design policy, with a new national goal and a more comprehensive approach: "Architecture and design will help to create a sustainable, equitable and less segregated society with carefully designed living environments in which everyone is well placed to influence the development of their shared environment". The policy also directly references the international perspective within "global sustainable development" and includes generally sustainability and quality, not being subject to "short-term financial considerations". Additional competences for the Swedish Housing Agency make it responsible for coordination, supporting the development of capacities, and promote the implementation, on the national, regional, and local level on the designed living environment, which includes developing "tools that municipalities are calling for and document good examples". One example would be to produce guidance for municipalities for "qualitative, accessible and sustainable design of the physical environment of schools and preschools". Other aspects of the policy include sustainable transport (using proximity in urban environments) and the use of sustainable materials. The mandate of the "Council for Sustainable Cities"—created in 2017—has been expanded until the end of 2030 (Hållbarstad 2023), which aims to facilitate cooperation between authorities contributing to the implementation of SDG 11, provide up to date information on knowledge support and funding opportunities, and support particularly municipalities affected by major industrial establishments and industrial expansions.

In addition to the legislative acts described, additional policy instruments and tools complement the national level support for municipalities. These include, for instance:

- A Strategy for Liveable Cities (*Politik för levande städer*, Skr. 2017/18:230) has also been adopted in early 2018 and serves as Sweden's first national urban development strategy. It aims in particular to implement SDG 11 (on sustainable cities) and provide the conditions for municipalities to develop green, healthy, and safe cities.

- The MayorsIndicators tool (MayorsIndicators 2023) was developed to monitor and compare the SDG indicator implementation between Finnish, Swedish, and UK municipalities (see above for a short overview in the context of Finland). Luleå and Kiruna both rank in the highest category with scores of 1065.24 and 1066.94, respectively. More detailed results are available only to the network's members.

The strong standing of Swedish municipalities in the constitutional and national legal framework gives them a large degree of discretion over their local affairs, including their urban planning. Sustainable development in its various notions (planning, design, transport) has been increasingly mainstreamed with policies on the national level to assist them and coordinate the implementation better. One piece of feedback from our interviews in Luleå included that the city is orientating itself at the national level, but also need to ensure that their views are brought to the national and EU level to say what they need (elected official, Luleå, 30.09.2022). While this subsection can only give a limited insight in a few key elements, the wider context of policies relating to urban development, however, has been assessed as "highly fragmented and relatively powerless" (Lidström/Hertting 2021, p. 281) with a focus on enhancing "the implementation of other established sectoral policy goals through improved coordination rather than creating a substantive policy goal in itself" (Lidström/Hertting 2021). Another piece of feedback from the interviews was that the Swedish planning law does not specify how exactly to involve citizens and stakeholders (interview, researcher, Luleå,). This critical assessment highlights the strong position of local self-government which makes a look at the local context of implementation (see below Section 4.3 and Chapter 5) even more important.

4.4 Approaches to Implementation by Cities and Translation to Local Policies

The second step in translating the international instruments' approaches takes place on the local level. The cities build on the national constitutional and legal framework and use programmes and tools from the national level. This section gives a short overview of the local discourses on urban sustainable development in our seven case study cities including responses from the interviews (see Chapters 1 to 3 for the approach and methods). Maintaining the structure from the previous Section 4.3, the case study cities are described in the order of their country.

4.4.1 Rovaniemi and Kolari/Finland

Under the Sustainable City Programme (see above Section 4.3.1), Rovaniemi participated in a project on the cross-effects of sustainability actions, with a particular focus of the municipality on the service network planning. The project aimed to support the participating five cities to identify economic, environmental, social and cultural cross-effects of activities promoting sustainable development and strengthening these effects (Gaia 2023, p. 26ff.): The project report's recommendations include inter alia the improved use of existing fora and information, raising awareness and improving the knowledge base within the municipality, including by utilising expert consultations and reports to increase inputs from the outside. Other recommendations included improving functional practices for the participation of municipal residents and other groups to allow co-development (as opposed to only specific entry points for participation), and more comprehensive impact assessments in the municipality's decision-making process to broaden the view on potential effects of decisions.

Overarchingly, this ties in with the feedback we received from our interviewees. Regarding the implementation of urban sustainable development, there can be different expectations from the individual, municipal, and national level while acknowledging interdependencies between these levels: one of our interviewees, for instance, underlined the importance of social factors also in sustainable development strategies and felt that people from the Southern part of the country, including the capital Helsinki "quite often [...] see Lapland like kind of a big nature reserve. And that's not really sustainable from the local point of view" (business representative, Rovaniemi, 18.02.2022). Other interviewees shared that Finnish cities have sustainable development strategies that influence regional development and require financing for their implementation. Influencing the national and EU policy would be key to help the sustainable development on the local level in this regard, including the availability of taxes to the local level (elected official, Rovaniemi, 24.11.2022, see also above Section 4.3.1).

In Rovaniemi, the local context for sustainable urban development that our interviewees highlighted included aspects of tourism, transportation, recycling, and green urban infrastructure as well as renewable energy. Some responses indicated that the municipal policy- and decision-making would—despite the general sustainability angle—fall short from that ambition, be it transporting biowaste to Oulu instead of using it in the city, still using much more concrete instead of more sustainable wood for construction, lacking more green urban planning in the city centre and avoiding additional wind farms that could interfere with touristic goals/views (elected official, Rovaniemi, 02.02.2022).

In Kolari, the interviewees underlined the mismatch between individual or family level planning for sustainability and municipal planning periods, with one interviewee stating that "municipal and regional levels are far from ordinary people and their priorities" (researcher, Kolari, 28.01.2022). This gets more complicated as even in this comparably small municipality, the interviewee highlighted that "we are not [a] homogeneous population" (researcher). Another interviewee highlighted regarding the localisation of the sustainable urban development should be "based on renewable and intensifying demographic development, the constant renewal of infrastructure and a responsible approach to the specificities of the regions" (business representative, Kolari, 04.03.2022).

In the local context of Kolari, the main discussion on urban development that overshadows all others, relates to the big two sources of revenue in the city: resource extraction and tourism. Based on the responses in the interviews, the discussion is both very decidedly partisan (clear pro or clear contra in the discussion) yet very nuanced in the respective reasoning, which seems difficult to reconcile in local planning. Both industries provide (and have been providing) necessary job opportunities and contribute to diversifying the economic development in the region (business representative, Kolari, 04.03.2022). They also require detailed planning including social and ecological sustainability, as well as a lot of investment, and are affected by transport connections, energy supply, and energy prices. Additional complications stem from the importance of reindeer herding for local and Indigenous communities which can use tourism as a second source of income, but are often negatively affected by the use of space in their grazing areas, for example, also by windmills (business representative, Kolari, 04.03.2022). The discussion on taxes for the local authorities is also relevant here, as mine workers tend to fly in and out so that tax revenue is lost to the local level that has to deal with the (environmental and social) impacts of the resource extraction. In both industries, the discussions on sustainable approaches and their limits are ongoing: one of the newest additions in the Northern part of the Kolari municipality is the "Ylläs tourist area master planning and business concept" that aims to further develop the already biggest skiing resort in Finland, which is also located in a national park, planning to add 10,000 beds, about 80% of the existing capacity, and expanding the skiing slope capacity by 60% (Ylläs 2023). With regard to the mining activities, local citizens continue to pursue legal remedies against the licensing (business representative, Kolari, 07.03.2022).

4.4.2 Akureyri/Iceland

Akureyri has established a master plan (2018–2030) for the merged municipalities of Akureyri with the islands Hrísey and Grímsey, which includes the main planning document, (binding) land use maps, and an environmental report (Akureyri 2018). It is accompanied by several thematic plans for clarification and particularly protected houses. Sustainable development is reiterated several times as a guiding principle, including when planning settlements (Akureyri 2018, p. 24), which includes for the implementation that "environmental aspects, such as geology and landscape, must be accounted for in the planning and implementation plans as appropriate". In the context of compatibility with the national planning policy, the masterplan responds to the policy's aim that "structure, including transport" should be "sustainable backbones of the relevant local community"; urban planning should be recognised as contributing "to the sustainable development of urban areas with dense, continuous settlements, the reorganization of underutilized areas and the strengthening of local communities" (Akureyri 2018, p. 103). The master plan aligns with the national policy regarding goals for integrated residential patterns and continuous settlements, the quality of the residential environment and natural environment, density of settlements, growth limits, diversification of the economy, transport, [...] together with environmental issues in general (Akureyri 2018, p. 103).

The coverage of local plans in Akureyri can be seen on a local map service (Map. is 2023). Additional plans by the municipality that are affected include its housing plan, as well as its school, employment, tourism, cultural, welfare, equality, architectural, environmental, and transport policies (Akureyri 2018, p. 106).

Our interviewees indicate that the discussion of urbanisation in the planning is a rather recent development, as previous regional development plans focused on the countryside and aimed at avoiding depopulation. As one interviewee put it:

> We haven't really talked about it [scil.: urbanism] because we are an urban state, but rural at heart because we think we are still living in the countryside, but we're actually living in cities and we haven't really noticed it just happened.
>
> (researcher, Akureyri, 27.05.2022)

Also, in the context of the small size of Akureyri (under 20,000 inhabitants), an interviewee underlined the importance of higher citizen involvement, but highlighted at the same time the difficulties to engage the working demographic aged 30–50 due to their time constraints (administration, Akureyri, 09.05.2023).

Various economic developments in Akureyri have conflicting goals, for example, growing business in the tourism sector, industry, or fisheries can come to the expanse of use of space, energy use, and environmental integrity. The tourism sector in the city grew of particular importance in recent years as it also took up local tourism from the Reykjavik region during the COVID-19 pandemic. The expansion of cruise ship tourism also contributes to the city's revenues.

Our interviewees added that the local discourse on sustainable urban development covers also local recycling, more sustainable transport—including changing the modal shift away from personal car usage, and methane-run buses—and energy planning. The energy discussion comprises the growing demand of local industries as well as the grid planning, which triggers local supply side investments in new power plants (resident, Akureyri, 26.04.2023).

4.4.3 Tromsø/Norway

Tromsø has been of academic interest for urban planning for over two decades, and the "Tromsø experiment" took already place in 2005–2006 (Nyseth 2011): this approach involved a network of local actors ("City Development Year Committee") that organised inclusive activities over a whole year in which the process was out of the city administration's hands. Deducing from our interviews, the sustainable urban development discourses that currently take place include some sceptical views regarding the full implementation of the SDGs and their indicators. One interviewee from the administration said that they "are not confident that [we] have found all the areas where we have to work with in order to deliver on all three dimensions" (elected official, Tromsø, 10.03.2023). The main policy in Tromsø since the 1990s had been the compact urban development which led to a densified city centre, but leaves the periphery behind in terms of transport connections (researcher, Tromsø, 08.03.2023). In the context of transport, several interviewees highlighted also the discussions on the pricing of toll roads, and as one interviewee phrased it: "So we are talking a lot about how we can include being green and build the city up, but also be out of the city because not everybody wants to be in the city" (elected official, Tromsø, 10.03.2023).

According to interviewees from the political level and businesses, the participation is limited to a group of citizens that are very active while the wider public would not take up current opportunities of involvement, also lacking additional outreach by the city (politician, Tromsø, 20.03.2023 and business representative, Tromsø, 25.04.2023). In this context, a landmark ruling of the Norwegian Supreme Court on the rights of Indigenous Peoples that found the construction of windmills to be illegal is also likely to shape the discussion on the local level in years to come (administration, resident, Tromsø, 24.03.2023), trying to reconcile the need for capacities and possibilities to develop sustainably (e.g., renewable energies) with the use of space and local rights. Finally, according to several interviewees, also the shortage of affordable housing is an ongoing problem (see also Section 4.1.1, Norway VNR 2021), which is partly linked to increasing tourism and private rentals (researcher and resident, Tromsø, 15.03.2023).

4.4.4 Luleå and Kiruna/Sweden

Luleå is part of the association of Swedish "Eco-Municipalities" (Sekom), which provides a network to exchange good examples and ideas. Participating municipalities need to adhere to four sustainability criteria, including reducing the extraction of natural resources, reducing the burden on nature, not using more natural resources that can be reproduced, and creating a society in which all people can meet their needs (City of Luleå 2024). The association follows up with its member cities on a number of indicators on urban sustainability, including for instance greenhouse gas emissions, modes of transport, amount of waste and recycling, renewable energy generation and fuels, as well as certified schools and kindergartens (Sekom 2024).

Luleå also updated its "Vision 2040" document in September 2021, adjusting the timeline of the previous Vision 2050 document, which had been already approved in 2008 (City of Luleå 2021). Its three main pillars are social, economic and ecological sustainability. While the document lays out a direction for future policies, it does not constitute a legal planning document. The Vision 2040 has been approved by the municipal council and involved citizens and businesses in the development. The directions of the vision were adopted by the Municipal Council in 2011 and include standing for openness and diversity, boosting identity as a coastal city all year round, betting on a leading Northern region, and building for the future. The statements of intent associated with these directions (City of Luleå 2021, pp. 21f.), include easier accessible participation for everyone (under "openness and diversity"), planning for a mixed, dense, and attractive city, and greatly increase the share of local journeys by public transport, bicycle and on foot (under "building for the future").

In line with these statements of intent, our interviewees mentioned particularly the exchanges on mobility, transport, and infrastructure, as well as high tech industries, start-ups and (renewable) energy investments (elected official, Luleå, 08.05.2023, business representative, Luleå, 01.04.2022, researcher, Luleå, 20.01.2022). One interviewee from the City Council stated that Luleå wants to develop a common vision on transport and infrastructure with another municipality, having ownership both, of the developing networks and the vision's content (elected official, Luleå, 08.05.2023).

In Kiruna, the transition of the city to allow for its iron mine to expand is the overarching topic of discussions on urban development, and even the city itself uses the moniker "a city in motion" (Kiruna 2024).

From several of our interviewees, we heard in the context of implementation that social and cultural dimensions of transformations would be less discussed and no broader discussion held on how the land is used (researcher and representative of civil society organisation, Kiruna, 17.01.2022). Also, the industry—in this example the mining operator Luossavaara-Kiirunavaara Aktiebolag (LKAB), which led the re-planning of the city—would shape how citizens are supposed to think about sustainable urban development (researcher, Kiruna, 20.01.2022). Sustainability would be too often seen by governments as a merely technical problem for which there are technical solutions (researcher and representative of civil society organisation, Kiruna, 17.01.2022) and continuing growth of the mine would also not lead to more jobs (researcher, Kiruna, 20.01.2022). Reconciling the trifecta of national interests in minerals, local nature, and Sámi rights would be difficult on the local level and the state would not do a lot in this context (researcher, Kiruna, 20.01.2022).

4.4.5 Nuuk/Greenland

In Nuuk, the local context for sustainable urban development is drawn between the local needs of the municipality and its citizens, the interests of the national government (Nuuk being the capital), and the wider international implications of Greenland's relationship with the Kingdom of Denmark.

For instance, one of our interviewees stated that "being environmental is not our first priority, but being independent [is]" (representative of civil society organisations and resident, Nuuk, 29.04.2022). The local understanding of sustainable development also follows a different focus compared to the other case study cities. For instance, eating locally means eating a lot of meat and seafood; limiting the use of cars takes place through the very limited and aging network of roads; not being able to leave the country, or even get around within the rest of Sermersooq municipality, without flights (researcher, Nuuk, 15.07.2022). The discourse on the development of utilities addresses challenges to keep up with growing demands by a growing population. The discourse on sustainable international exchange is linked to tourism: The development of tourism is linked to the expansion of transport facilities, including the airport, and an expanded harbour. At the same time, Nuuk claims to want to attract the "right tourists", while at the same time hoping to increase quality of life in the city itself with the expanded airport as well (representative of the administration, Nuuk, 04.04.2022). Using the advertising of sustainable tourism in this remote region includes a seeming paradox and a local promise that local initiatives such as "Visit Greenland" and "Colourful Nuuk" try to capitalise on (researcher, Nuuk, 04.08.2022).

4.5 Conclusions

This chapter shows a diverse landscape of governance on the international, the national, and the municipal/city level. The chapter's findings reflect the processes and factors of our conceptual model as presented in Chapter 2, and our conceptual model builds on three factors: (1) actors and their relationships, (2) institutions and their set-ups, and (3) political priorities. All three factors are affected by—and in return affect themselves—the underlying driver of imaginaries. The respective interlinkages can be found on all three governance levels covered in this chapter. The second driver in our model—cooperation—is particularly prominent on the international level, and so far only to a rather limited extent on the local level. This will be further explored in Chapters 5–7.

Firstly, the key international instruments for sustainable urban development in the European Arctic (see Section 4.1), including the 2030 Agenda, the New Urban Agenda, and the Paris Agreement included very different actors in their design. Both the 2030 Agenda and in particular the New Urban Agenda took a much broader approach to involve cities and municipalities and their networks in the discussions and creation, which is also reflected in the language addressing the local level, including (self) commitments, and the creating of indicators for the implementation, as under the 2030 Agenda process. The Paris Agreement, in contrast, focuses with its approach and commitments on state Parties with cities only being affected as sub-national entities in the implementation. Despite the different approaches of involving cities in the design of these three international instruments, it could be argued that the institutional setup of international instruments designed in intergovernmental processes clearly favours the political priorities and role of states' governments (and their representatives), which indicates that their imaginaries are being largely reflected with only very limited direct input by local governance actors.

Secondly, the national level (Section 4.2) provides the links between the international and local level via general constitutional provisions and specific legislative acts, such as the respective local government acts in four out of the five states (Greenland having an exceptional setup and no constitution), as well as sectoral legislation, for instance on building and planning as well as environmental impact assessments. Additional national policies supplement this governance level, aiming to provide political guidance together with financial support arrangements to strengthen local cooperation and coherence in their approaches to implementation. On the national level, the similarities of actors (national legislators), the institutional setups (constitutional framework, filled with general and specific legislation, as well as policies), and political priorities (delegating local decisions on sustainability to the local level) are very prominent. Also here, the national level's—that is, respective government's—imaginaries of the environment are the determining element, in several of our case studies being informed by exchanges with municipal associations and regional councils.

Thirdly, on the local level (Section 4.3), our case study cities generally have the competences to take and implement key decisions on their local approaches to sustainable urban development. This includes areas such as urban and land-use planning, mobility, energy supply, and overall well-being of its citizens. The overview on the Nordic Model (see Chapter 5) further elaborates on the implementation of these competences. While the key actors on the local level are the city administrations, including local businesses and civil society in their decision-making, their capacities for implementing sustainable urban development are determined by their respective institutional setup. From the feedback of our interviewees, there seems to be a substantial disconnect between the expectation for the local level's ambition and its personal and financial capacities to deliver. This leaves very limited leeway to implement local political priorities and in turn also limits the involvement of local imaginaries of the environment. As several of our case study cities attempt to cooperate with other Arctic cities, for example, via the Arctic Mayors Forum (AMF), which could improve the impact of their local imaginaries (see Chapter 8), here too their capacities are limited.

Acknowledgements

This chapter was written by Arne Riedel and reviewed by Jacqueline Götze, Michał Łuszczuk, Katarzyna Radzik-Maruszak, Dorothea Wehrmann and an external expert. All authors of this book read and accepted the content of this chapter.

Note

1 Only some of these documents are available in English, and most had to be machine translated (via Google Translate and Microsoft Translator, between October 2023 and February 2024), so English quotes can only be approximate to the original text and should not be taken as an official translation.

References

2030 Agenda 2015. Transforming our world: The 2030 Agenda for Sustainable Development. UN General Assembly. Available from: https://www.refworld.org/docid/57b6e3e44.html [Accessed 07 July 2023].

Akureyri 2018. Master plan for the municipality of Akureyri. Available from: https://www.akureyri.is/static/files/Skipulagsdeild/Adalskipulagid/ASAK_2018_2030/ASAK_breyting/augl/adalskipulag-akureyrar-2018-2030-greinargerd-i-gildi_med-breytingum-2019-2020-og-2021-okt22.pdf [Accessed 22 December 2023].

Alþingi 2023. Law database of the Icelandic Parliament (Alþingi). Available from: https://www.althingi.is/lagasafn/ [Accessed 22 December 2023].

City of Luleå 2024. Ekokommun. Available from: https://www.lulea.se/kommun--politik/sa-arbetar-vi-med/ekokommun.html [Accessed 01 July 2024].

Finland VNR 2020. Voluntary national review 2020 —Finland. Government of Finland—Publications of the Prime Minister's Office 2020:8. Report on the implementation of the 2030 Agenda for Sustainable Development. Available from: https://sustainabledevelopment.un.org/content/documents/26265VNR_Report_Finland_2020.pdf [Accessed 2 December 2024]

Gaia 2023. Report "Kaupunkien ja kuntien kestävyystyön ristikkäisvaikutukset" ("The cross-effects of the sustainability work of cities and municipalities") 2023. Available from: https://kestavakaupunki.fi/documents/100251420/0/Kest%C3%A4vyystoimien+ristikk%C3%A4isvaikutusten+arviointi.pdf/dd2cf8c8-bc6f-109e-c00d 9d5f0f2ca22e/Kest%C3%A4vyystoimien+ristikk%C3%A4isvaikutusten+arviointi.pdf?t=1701780182816 [Accessed 26 January 2024].

Hållbarstad 2023. Website. Council for sustainable cities. Available from: https://www.hallbarstad.se [Accessed 22 December 2023].

Iceland VNR 2023. Voluntary national review—Iceland 2023. Government of Iceland—Prime Minister's Office. Report on the implementation of the 2030 Agenda for Sustainable Development. Available from: https://hlpf.un.org/sites/default/files/vnrs/2023/VNR%202023%20Iceland%20Report.pdf [Accessed 22 December 2023].

Iceland National Planning Agency 2016. Landsskipulagsstefna 2015–2026 ásamt greinargerð, National Planning Policy. Available from: https://www.landsskipulag.is/media/pdf-skjol/Landsskipulagsstefna2015-2026_asamt_greinargerd.pdf [Accessed 22 December 2023]. The website of the National planning Agency includes additional documents related to the policy and its process: https://www.landsskipulag.is/gildandi-stefna/skjol/ [Accessed 22 December 2023].

Kestävä kaupunki 2023a. Website. Sustainable City—Yhteistyötahot (Collaboration Partners). Available from: https://kestavakaupunki.fi/yhteistyotahot [Accessed 26 January 2024].

Kestävä kaupunki 2023b. Website. Sustainable city- themes. Available fom: https://kestavakaupunki.fi/en/themes [Accessed 26 January 2024].

Kiruna 2024. A city in motion. Available from: https://kirunalapland.se/en/activities/city-in-motion/ [Accessed 01 July 2024].

Lidström/Hertting 2021. Limited, fragmented and powerless: national urban policies in Sweden. In: Zimmerman, Karsten & Fedeli, Valeria (ed.), *A Modern guide to National Urban Policies in Europe* (pp. 268-283). Cheltenham: Edward Elgar Publishing, Elgar Modern Guides. Available from https://doi.org/10.4337/9781839109058.00017 [Accessed 02 December 2024].

Map.is 2023. Akureyrabær (City of Akureyri), Loftmyndir ehf (data provider), GIS mapping of plans of Akureyri and surroundings. Available from: https://www.map.is/akureyri/# [Accessed 22 December 2023].

MayorsIndicators 2023. Website. MayorsIndicators tool. Available from: https://mayorsindicators.com/index.cfm [Accessed 26 January 2024].

Norway VNR 2021. Voluntary national review 2021—Norway. Report on the implementation of the 2030 Agenda for Sustainable Development. Norwegian Ministry of Local Government and Modernisation and Norwegian Ministry of Foreign Affairs. Available from: https://hlpf.un.org/sites/default/files/vnrs/2021/28233Voluntary_National_Review_2021_Norway.pdf [Accessed 22 December 2023].

Norwegian Ministry of Local Government and Regional Development 2017. Urban sustainability and rural strength—in brief. Norwegian Ministry of Local Government and Regional Development. Meld. St. 18 2016–2017 Report to the storting (white paper). Available from: https://www.regjeringen.no/en/dokumenter/meld.-st.-18-20162017/id2539348/ [Accessed 26 January 2024].

Norwegian Ministry of Local Government and Regional Development 2023. A good life throughout Norway—regional and rural policy for the future. Norwegian Ministry of Local Government and Regional Development. Meld. St. 27 (2022–2023) Report to the storting (white paper). Available from: https://www.regjeringen.no/en/dokumenter/meld.-st.-27-20222023/id2985545/ [Accessed 26 January 2024].

OECD (2023). OECD Regional Outlook 2023 – Country Profiles: Iceland. Available from: https://oecd-cfe-eds.github.io/ro2023-country-pages/tl0-isl.html#recent-policy-developments [Accessed 02 December 2024].

Paris Agreement 2015. Paris Agreement to the United Nations Framework Convention on climate change, Dec. 12, 2015, T.I.A.S. No. 16–1104.

Policy for Designed Living Environment 2018. Policy for designed living environment. Government offices of Sweden—Ministry of culture. Abridged information document for public information. Available from: https://www.government.se/contentassets/c008469d86b848f3918a1efcd7d7fb2f/policy-for-designed-living-environment.pdf [Accessed 22 December 2023]. The full text of Bill 2017/18:110 is available from the Sveriges Riksdag (Swedish Parliament) Available from: https://www.riksdagen.se/sv/dokument-och-lagar/dokument/betankande/politik-for-gestaltad-livsmiljo_h501ckru1/html/#_Toc514747660 [Accessed 22 December 2023].

Sekom 2024. Gröna nyckeltal. Available from: https://sekom.miljobarometern.se/lulea [Accessed 01 July 2024].

SFS. Svensk författningssamling, Official database for the Swedish Code of Statues. Available from: https://svenskforfattningssamling.se/ [Accessed 02 December 2024].

Sweden VNR 2021. Voluntary National Review 2021—Sweden. Report on the implementation of the 2030 Agenda for Sustainable Development. Government offices of Sweden. Available from: https://hlpf.un.org/sites/default/files/vnrs/2021/279582021_VNR_Report_Sweden.pdf [Accessed 22 December 2023].

UN Conference on Housing and Sustainable Urban Development 2017. New Urban Agenda. publication by the Habitat III Secretariat. Available from: https://www.habitat3.org/the-new-urban-agenda [Accessed 26 January 2024].

UNFCCC 1992. United Nations Framework Convention on climate change, 9 May 1992, S. Treaty Doc No. 102–38, 1771 U.N.T.S. 107

Ylläs 2023. Sport resort: Major new development plan in Ylläs. Available from: https://www.lapland.fi/business/sport-resort-yllas-project [Accessed 01 July 2024].

5 The Nordic Model Puzzle

Unpacking Participation and Local Approaches to Sustainable Urban Development

5.1 Introduction

The cities under investigation in the European Arctic are located in different countries (see Chapter 1). They have diverse histories and are confronted with varying opportunities and challenges when it comes to sustainable urban development. Furthermore, as we show in this chapter, the cities apply different participatory approaches deriving from their interpretation of legal frameworks as well as from cooperation formats. Despite these differences, they also share similarities in terms of governance models, economic development, climate change vulnerability, and future prospects. This chapter is on the one hand an evaluation of the effectiveness of practices of Nordic Model at the local level and on the other a reflection on participation in the analysed municipalities.

This chapter is structured as follows. First, we introduce the Nordic Model that constitutes a ground for operation of national and local governments in Nordic countries. Then we focus on governance structures in and around the analysed cities to illustrate the model's practicalities. In the next section, we pay attention to citizen participation, including the ability of citizens to shape urban policies in Nordic countries. Finally, we present research findings from our case-study municipalities and show a gap between the assumptions of the Nordic Model and the governance practices in Arctic municipalities. Although the Nordic Model formally presupposes trust, inclusion, equality, and transparency in decision-making processes, our research findings reveal an imbalance among actors. This is primarily due to the knowledge(s) and resources they possess.[1]

5.2 The Nordic Model of Governance

Denmark,[1] Iceland, Finland, Norway, and Sweden collectively embody a profound cultural legacy, underpinned by values such as democracy, justice, equality, and transparency. These nations implement congruent public policies, as noted by Blossing *et al.* (2014) and Antikainen (2006). Renowned for their robust democratic frameworks, the countries also share similar approaches in structuring both central and local governance, as well as policy- and decision-making mechanisms. Furthermore, the Nordic countries are characterised by their willingness to share

DOI: 10.4324/9781003284864-5

knowledge and engage in collaborative endeavours across various levels—local, national, and international. This synergy of a potent state apparatus, effective legal systems, and accountable democracies is identified by researchers, including Fukuyama (2014), as a key ingredient for exemplary governance.

Comparative studies, such as those by Haveri (2015), consistently rank these Nordic states highly, emphasising their unique ability to merge economic efficiency and growth with labour market stability, equitable income distribution, and social unity. Johansson (2022) further elaborates on the region's socio-political landscape, highlighting the enduring influence of social democracy. This is evident in the dominance of social democratic parties, influential labour movements, high unionisation rates, and extensive involvement of social partners in wage discussions and determinations. Furthermore, the countries are noted for their universal social welfare policies, encompassing comprehensive social protection systems and services that cater to the entire populace. Contrasting with many European counterparts, Nordic public administrations are distinguished by minimal corruption and high public trust, as noted by Andreasson (2017). These distinctive traits are collectively referred to as the "Nordic Model", also known as the "Nordic welfare model", the "Scandinavian model of welfare", or the "social democratic welfare state regime", terms echoed by scholars such as Johansson (2022), Schrage and Kjærås (2022), and Esping-Andersen (1990). In the following sections, with a main focus on the local level, we delve into various aspects of this model, exploring its potential benefits and challenges to aligning policies on urban development.

To enhance the introduction of the model's components, we begin with an examination of the organisational structure at both central and local levels. Each country is characterised as a parliamentary state, featuring multi-party systems and coalition governments. These governments are known for their consensual approach to policy-making, as noted by Greve *et al.* (2016a) and Painter and Peters (2010). Legislative power is vested in unicameral parliaments, while the executive branch is led by prime ministers. The central administrative machinery operates on a professional and non-politicised basis, with a significant role played by central agencies in the governance process. Nordic countries exhibit a high degree of decentralisation, assigning substantial importance to local governments (Baldersheim *et al.* 2019, see also Chapter 4). This decentralisation is not a recent trend but a historical characteristic of these countries/societies, as Johansson (2022) outlines. Some scholars even propose that the essence of modern Nordic countries can be traced to the local origins of social policy developments from the 20th century.

At the local level, there is a bifurcated structure—with the exception of Iceland—comprising regions and municipalities (refer to Table 4.1). Regions typically assume coordination responsibilities, whereas municipalities deliver a broad spectrum of services. These services include healthcare, education, childcare, elderly care, spatial planning, and environmental management. Consequently, municipalities are often termed "service providers" or "facilitators of social welfare provision" (Stoker 2011, p. 15). In this vein, Trydegård and Thorslund (2010) emphasise that Nordic countries adopt a "welfare municipalism" model. This model equips local governments not only with extensive responsibilities but also

Table 5.1 Territorial structure of Nordic local governments

Number of units	*Denmark*	*Finland*	*Greenland*	*Iceland*	*Norway*	*Sweden*
Municipalities	98	309	5	64	357	290
Regions	5	19	–	–	15	21

Source: Own preparation based on: Denmark. Unitary Country: https://www.sng-wofi.org/country-profiles/denmark.html (accessed 17 September 2023); Finland. Unitary Country: https://www.oecd.org/regional/regional-policy/profile-Finland.pdf (accessed 17 September 2023); Record Population Increase in Iceland: https://www.icelandreview.com/society/record-population-increase-in-iceland/ (accessed 17 September 2023); Norwegian Local Governments Sector Outlook 2023: https://www.scopegroup.com/dam/jcr:ff37f1de-3531-4de3-9624-f9bc97d605af/Scope%20Ratings_Norwegian%20local%20governments_Sector%20Outlook%202023_Final.pdf (accessed 22 May 2024); Counties and municipalities in Sweden: https://www.scb.se/en/finding-statistics/regional-statistics/regional-divisions/counties-and-municipalities/ (accessed 22 May 24).

with the capabilities to fulfil them effectively (Anttiroiko and Haveri 2022, p. 65). In this context, some scholars argue that Nordic countries can be seen as nationalised local governments, emphasising the local roots of 20th-century social policy developments (Johansson 2022, p. 417, see also Wennemo 2014). The structure of Nordic local governments is displayed in Table 5.1.

Nordic local governments enjoy a considerable high degree of autonomy regarding the organisation of local service provision (Greve *et al.* 2016) and the structural arrangement of local authorities. Nevertheless, the majority of local governments exhibit a preference for collective decision-making bodies, such as councils and executive boards, over strong monocratic leadership. This considerable autonomy is linked to the "free municipalities" experiment conducted in several Nordic countries during the 1980s. This initiative allowed basic administrative units—municipalities—to expand their competencies and autonomy. The programme was successfully launched in Sweden (1984), Denmark (1985), Norway (1986), and Finland (1989) (see Baldersheim 1994).

The Nordic countries have also implemented similar political and social reforms. They have embraced substantial changes in service delivery, drawing inspiration from New Public Management and New Public Governance principles (Haveri 2015, Greve *et al.* 2016). In an ongoing process that started in the 1950s, Finland, Norway, and Sweden have pursued municipal amalgamations. Since then, Finland, Norway, and Sweden have significantly reduced their number of municipalities. Currently, territorial reforms are noticeable in Denmark, Greenland, Finland, and Iceland (Eythórsson *et al.* 2018, Karlsson and Eythórsson 2022). In Denmark, the goal was to augment the roles of municipalities and regions by consolidating them into larger entities (Blom-Hansen and Heeager 2010). Today, Danish municipalities rank among the largest in the Nordic region in terms of population (Johansson 2022, p. 418). Similar reform was introduced in Greenland in 2009 when 18 municipalities were reduced to 4 (see Hansen 2015, Eythórsson *et al.* 2015). In Finland, the reform process began with enforced amalgamations (Vakkala *et al.* 2021); however, it now proceeds in a more measured and systematic fashion. A yearly reduction in the number of municipalities has been observed. Lastly, as of 1 January 2024, Iceland had 64 municipalities, which represents a decrease of five due to consolidation efforts (Record Population Increase in Iceland 2023; see Table 4.1). In many cases, amalgamation reforms have imposed stronger inter-municipal cooperation (Eythórsson *et al.* 2017).

Additionally, it should be underlined that Nordic countries exhibit a cohesive approach towards citizen engagement and inclusion. Public participation is a fundamental component of the Nordic Model. Local governments, adhering to network governance principles—which form the cornerstone of the decision-making process and are theoretically inclusive and built on mutual trust (Andreasson 2017)—are obliged to integrate actors from the public, private, and societal sectors into their decision-making processes (Torfing *et al.* 2012, Sorensen 2022). However, despite the shift towards participatory and deliberative democracy in many European nations, Nordic states continue to prioritise representative democracy at the core of decision-making. While local-level trials with more direct participatory methods, such as binding referenda and participatory budgeting, have been implemented, these democratic innovations typically serve as enhancements to established representative mechanisms. Empirical research suggests that, across all Nordic countries, professional expertise is often valued above the lay knowledge of "ordinary citizens" (Radzik-Maruszak 2019). The populace is frequently viewed more as recipients of services rather than as active stakeholders in policy-making, which, in turn, curtails social inclusivity in local strategic planning for urban development.

5.3 Governance Structures and Local Approaches to Sustainable Urban Development in the Cities Investigated

As introduced in the initial section on the Nordic Model, the Nordic states exhibit similar governance structures at the local level. This encompasses the organisation of local authorities and their interactions with other stakeholders present at the local level. Regarding the former, it is noteworthy that Nordic states favour collegiate bodies and collaborative decision-making over individualistic leadership. Haveri (2015, p. 139) suggests that this may stem from the historical development of local government institutions in the Nordic countries, which were influenced by a free peasant society. This societal structure fostered the evolution of autonomous municipalities, with a focus on welfare and a tradition of collective leadership. The historical ascendancy of peasant collectives nurtured the notion of entrusting responsibilities to group entities rather than to individual public officers. Furthermore, these societies were tasked with community and economic stewardship, necessitating consensus for decision-making and thereby reinforcing a culture of collaboration and consensual policy (Haveri 2015, p. 140, cf. Wetterberg 2004, p. 21). Currently, this legacy of extensive negotiations, cross-party alliances, and compromises remains a significant aspect of the political and administrative ethos in the Nordic countries, where pragmatism frequently takes precedence over ideological disparities (Haveri 2015, p. 140).

At present, power and responsibility within Nordic municipalities is delineated between political leadership and administrative management. The former is responsible for setting objectives and strategies, while the latter is charged with their execution, although the demarcation between these roles is occasionally ambiguous (Vakkala *et al.* 2021, cf. Leinonen 2012).[2] For instance, in Finland, New Public Management reforms instituted in the 1980s and 1990s augmented the influence and status of municipal administrators—particularly the municipal CEO—relative to elected officials (Mouritzen and Svara 2002). Nonetheless, in line with the principle of collective decision-making, the principal authority in Nordic municipalities remains with the local council. Across all Nordic municipalities, the council exhibits several uniform characteristics. First, the council always consists of an odd number of members. Second, each councillor is backed by a deputy from the same party list. Third, in contrast to, for instance, Central-Eastern European countries, a majority of councillors are affiliated with a political party. Goldsmith and Larsen (2004) note that in Finland, 95% of councillors are elected under a party banner. Additionally, significant roles within the council are fulfilled by boards or committees. These smaller entities, composed of councillors and external members such as experts and citizens, handle specific matters including infrastructure, planning, education, etc. There is greater variation in the organisation of executive power in Nordic municipalities. While the prevailing model is an executive committee (Norway, Sweden) or board (Iceland, Finland) appointed by the council, some countries have experimented with elected mayors. A prime example is Finland, where the revised Local Government Act (410/2015) introduced in 2015 presented new governance models intended to reinforce political leadership (see Chapter 4 for a general overview on Finland's and other Local Governance Acts). This reform permitted municipalities to determine the executive model that best suited them. As a result, seven municipalities opted to substitute the municipal manager with a mayor (selected from among the council members).[3] Nevertheless, in Finland, the balance of power between politicians and administrators still skews towards the latter (cf. Goldsmith and Larsen 2004, Vakkala *et al.* 2021).

Our case studies invariably illustrate a "conventional" configuration of local governance structures (see Table 5.2 for details). Yet, certain distinctive features have emerged. Specifically, in Kolari and Tromsø, we observed the election of independent, non-affiliated councillors. In Kolari, their emergence was linked to a contentious debate concerning the reopening of a mine, where it was perceived that the established political parties did not adequately represent the interests of those residents opposed to the initiative. In Tromsø, during the 2020–2023 term, four councillors were elected under the "People's Action No to More Road Tolls" banner, a movement borne out of opposition to toll fees instituted by municipal authorities.

The second level of local governance in most Nordic countries constitutes regions (see Table 5.1).[4] The regional government serves mainly as a coordinator of service provisions, however, presently following the complexity of governance structures they also perform a key role in national and international networking. The authority at that level is mainly divided between councils/assemblies and

Table 5.2 Organisation of local authorities in investigated case studies

Town/city	*Legislative body*	*Executive city*
Akureyri	The town council—11 members and alternates	Executive committee and CEO/Director*
Kolari	Town council—21 members	Executive board—7 members
Kiruna	City council—45 members	Executive board—15 members
Lulea	City council—61 members	Executive board—15 members
Nuuk	Municipal council (Kommunalbestyrelsen)—19 members	Mayor and deputy mayor
Rovaniemi	City council—51 members	Executive board—11 members
Tromsø	The municipal council—43 members	Executive commission—13 members

Source: Own preparation based on municipal webpages: Akureyri municipality 2024a: https://www.akureyri.is/en/administration/about-akureyri/town-council (accessed 31 May 2024); Kiruna municipality: https://kiruna.se/kommun--demokrati/kommunens-organisation.html (accessed 23 May 2024); Kolari municipality: https://www.kolari.fi/fi/kuntainfo/talous-ja-hallinto.html (accessed 31 May 24); Lulea municipality: https://www.lulea.se/kommun--politik/organisation/kommunfullmaktige.html; https://www.lulea.se/kommun--politik/organisation/kommunstyrelsen.html (accessed 23 May 2024); Sermersooq municipality: https://sermersooq.gl/da/committee/kommunalbestyrelsen/ (accessed 31 May 2024); City of Rovaniemi Administrative Regulations: https://www.rovaniemi.fi/loader.aspx?id=f51cc0ff-fbba-4173-ba03-510ad4ae3f39 (accessed 31 May 2024); Rovaniemi city board: https://www.rovaniemi.fi/Kaupunki-ja-paatoksenteko/Paatoksenteko/Kaupunginhallitus (accessed 31 May 2024); Tromsø City Council https://tromso.kommune.no/politikk/politisk-organisering/kommunestyret (aAccessed 31 May 2024); Tromsø executive commission https://tromso.kommune.no/politikk/politisk-organisering/formannskapet-og-hovedutvalgene (accessed 31 May 2024).

executive boards, which are often chaired by governors. Regarding the countries investigated, the regional level is present in Finland, Norway, and Sweden (see more Table 5.1). Our case cities are located in Lapland (Kolari, Rovaniemi), Finnmark (Tromsø), and Norrbotten Region (Kiruna, Luleå).

What is important, however, is the fact that our case cities—to a different extent—are active at different governance levels (see Chapter 7). In Finland, we analysed two cases—Rovaniemi and Kolari. The first constitutes important venue of collaboration regarding Arctic affairs, what can be illustrated *inter alia* by so called the "Rovaniemi Process". This initiative, launched by Finland in 1989, marked the beginning of more sustained intergovernmental Arctic cooperation (Nilson 1997). It was a response to a shift in geopolitical dynamics at the end of the Cold War and emphasised environmental protection, scientific research, and peaceful exploitation of natural resources. The process led to the establishment of two task forces, focusing on the Arctic environment and international law in the region. It also laid the groundwork for the creation of the Arctic Council (Koivurova and Hasanat 2009, Dodds and Woodward 2021). The city also hosts the International Secretariat of the University of the Arctic (UArctic): a network of universities, colleges, research institutes, and other organisations focused on (higher) education and research in the North. This initiative promotes cooperation in various fields related to the Arctic region (Hesseln *et al.* 2013). Additionally, Rovaniemi has contributed to the emergence of various Arctic dialogue forums, such as the Arctic Spirit Conferences, addressing the region's role as an epicentre of climate change and its growing geopolitical and economic significance, as well as the role of the Arctic's Indigenous. Presently, Rovaniemi has 16 twin cities; among them, the oldest partner municipality is Kiruna, with which a partner municipality agreement was signed as early as 1940 (International Rovaniemi 2024b). Rovaniemi terminated its twin city status with Murmansk a week after Russia's full-scale invasion of Ukraine in February 2022 (*The Barents Observer* 2022). In comparison, Kolari is a much smaller, sparsely populated municipality located near the Finnish and Swedish borders. The city is not very active for international formats. However, Kolari belongs to the Council of Torne Valley, a Swedish federation of municipalities and a transnational Nordic committee based in Haparanda (Tornio 2024).

In Sweden, both Kiruna and Luleå are involved in different cooperation formats. However, Kiruna as a smaller municipality with fewer staff has more limited capacity to engage at other governance levels and, for instance, in exchanges with other municipalities, although they desire to do so. Kiruna has several twin-city arrangements, for example, as mentioned before, with one of our other case study cities, Rovaniemi in Finland. However, our data from Kiruna does not indicate that these forms of cooperation are utilised for sharing experiences or knowledge(s) on sustainable urban development. In the case of Luleå, we noticed a different type of commitment to cooperation, first and foremost, through the municipality's and the mayor's personal engagement with the Arctic Mayor Forum (AMF). Furthermore, Luleå has several twin cities, including our case study city of Tromsø, Norway. Additionally, to this pan-Arctic focus of cooperation, Luleå is also engaged in European and Baltic cooperation, including the EU Covenant of Mayors for Climate & Energy or the Bothnian Arc association, which is open for municipalities, regional municipalities, and provinces located in the Bothnian Arc area in Finland and Sweden.

Tromsø has been labelled the "Gateway to the Arctic" and plays a significant role in political activities, particularly related to international Arctic affairs. Presently, Tromsø is a hub for Arctic governance and international cooperation, particularly in the context of environmental, geopolitical, and security issues in the Arctic region (see Chapter 1). The city's strategic location, being at a crossroads of the Norwegian, Finnish and Swedish Arctic and close to the Russian Arctic, underscores its significance in Arctic affairs. The city's role in hosting critical discussions and diplomatic engagements further strengthens its position as a key player in Arctic politics and international relations.

Akureyri, known as the *Capital of North Iceland*, has a university (Edvardsson 2014) and is involved in cooperation, notably through its twin-city relationships, for example, with Álasund, Lahti, Randers, and Västerås (Akureyri municipality 2024b). In November 2022, Akureyri debated its relations with its Russian twin city, Murmansk, and issued a statement condemning Russia's invasion of Ukraine. The town council proposed to end the friendship agreement (Nilsen 2022).

Nuuk is both the Greenlandic capital and the administrative seat of the large Sermersooq municipality, spanning also parts of Greenland's east coast. Under the Self-Rule Act of 2009, Greenland has authority over a number of policy fields (see Chapter 4), but inter alia foreign policy and security policy remain in the hands of

Denmark. Depending on the primary focus of international exchanges, the relevant competence can become a somewhat "grey zone" (Pram Gad 2022). The European Union's Commission has opened a representative office in Nuuk in March 2024 (EU's Arktiskontor i Grønland 2024). The city of Nuuk itself has several twin cities, including Reyjkavik, Iceland and Stockholm, Sweden. Nuuk has also hosted the "Arctic Circle"—Forum in 2022, as well as the EU Arctic Forum and Indigenous Peoples' Dialogue in 2023, and the municipality of Sermersooq is a member of the Arctic Mayors Forum (Gad 2022, EU's Arktiskontor i Grønland 2024).

5.4 Local Participation in the Nordic Countries

Public participation has been developed and studied for decades and various concepts were established (Kantola and Tuulentie 2020, Arnstein 1969, Beierle and Cayford 2002, White 2000). As described above, regarding participation, the Nordic states have certain features that make them unique. On the one hand, they all officially exhibit a distinctive approach to citizen engagement. These countries are often recognised for having the world's happiest citizens (Martel *et al.* 2020, Stende 2017) and for prioritising the quality of political institutions. In this regard, maximising citizen participation and representation in decision-making processes is believed to bolster the effectiveness of political institutions and strengthen public trust, a concept known as "Nordic exceptionalism" (Feeley and Langford 2022). The concept of network governance, which involves the inclusion of different actors from the public, private, and social sectors in decision-making, has gained traction (Löfgren and Ringholm 2009, Hall *et al.* 2009). On the other hand, research and practice suggest that representative democracy remains central to participation processes in the Nordic countries. There is a reluctance to adopt participatory or deliberative mechanisms for several reasons. First, the principle of social equality is paramount in these states, and there is a concern that participatory or deliberative methods may not afford equal voice to all social groups. Second, Nordic municipalities have been reoriented as "service machines" under the influence of the New Public Management paradigm (Radzik-Maruszak 2019, Nyholm 2006, Nyholm and Haveri 2009). That shift has led to an increased burden of responsibilities and susceptibility to central government interventions, which has constricted the scope of local democracy (Haveri 2015). Despite these challenges, there has been a renewed interest in new forms of participation. Yet, in many cases, citizens are regarded more as co-producers and co-creators of services than as co-decision-makers.

Presently, participation in the Nordic countries can manifest in both formal and informal ways (cf. Bherer *et al.* 2023, p. 1–2). Formal participation often involves institutional, top-down arrangements that are ingrained in the political and legal framework. Despite their variety and innovativeness, most of these mechanisms are criticised for being superficial, overly procedural and oriented towards public display. Such instruments include local elections and referenda, public consultations,[5] and participation in various social councils (e.g., for youth, senior citizens, citizens with disabilities, etc.; see Pawłowska *et al.* 2021). It is important to note that the deliberations of these bodies are generally advisory, as the ultimate decision-making power resides with elected regional or local governments. Concurrently, many Nordic local governments, particularly in urban areas, have begun experimenting with social innovations such as PB and "urban laboratories" (Smith 2009). In participatory budgeting (PB), residents are engaged by local governments to help determine public spending priorities, propose initiatives, and vote on which projects should be enacted by local authorities. PB is founded on a participatory decision-making process that is, in theory, deliberative, transparent, and inclusive (Lehtonen and Radzik-Maruszak 2024, p. 74). Urban laboratories, on the other hand, represent a compendium of methodologies devised for orchestrating urban experiments. These have evolved from purely technical exercises to more socially oriented endeavours that foster interaction and dialogue among different actors through various methods, such as charrettes, café dialogues, consultations, deliberative panels, workgroup weekends, forecasts, and artistic interventions (Nyseth *et al.* 2019, p. 8). Additionally, there is informal participation, which occurs outside officially sanctioned channels and offers an alternative to institutionalised forms that may be unwelcoming or unfamiliar to citizens. Examples include protests and social media communication.

In recent decades, Nordic countries have tried to include citizens' opinions into national and local policies. For this book particularly, their involvement in urban policy and sustainable urban development is important. According to Manzi and colleagues (2010), the latter concerns social inclusion and exclusion, feeling at home in the neighbourhood, citizen participation, empowerment and democratic governance, as well as the importance of integrated planning of housing, services, transportation, health, education, etc. Importantly Nordic countries have a long record of keeping this attitude (2010; see also Chapters 1 and 4). Manzi and colleagues (2010) explain that this is related to the build-up of the welfare state in the period between the 1940s and the 1970s. According to them, "[s]eeking social sustainability promotes new alliances and new ways of mobilizing resources through the new machineries of governance". This has opened the door to new alliances that imply new ways of solving problems, which have traditionally been the responsibility of established government structures. Creating new ways of participation and engaging citizens more directly in urban planning and development can be considered a good example in that context. Additionally, as Tunström (2019, p. 42) indicates, the involvement of the new groups and finding new ways of organising the planning process are seen as a remedy to the ills of previous eras. In this context "planning for people" has become a good motto for many Nordic cities (Participatory Urban Planning in Nordic Countries 2023, p. 9). This motto is, however, often an empty signifier. On the one hand, even though implementing participation formats is mandatory and spatial planning needs to follow different acts and entitlements, it is left with the municipality to design the processes and, by that, leaves much flexibility for not only how but also what kind of participation is actually implemented. As a result, the degree to which citizens and stakeholders are involved and what kind of participation is envisioned lies within the responsibility of the municipality. On the other hand, the scholarship indicates that municipal planners often prefer to provide excellent public service for people in building attractive cities rather than fully engage them in planning and perceive them as active actors of knowledge (Participatory Urban Planning in Nordic Countries 2023). This may be linked with the emphasis that is put in Nordic countries on output rather than input and throughput legitimacy (see more in Svels and Thuesen 2024). Therefore, even if the idea of public participation is high in the urban planning agenda in Nordic countries, it still suffers from kinds of "tokenism", as indicated by Sherry Arnstein's ladder of participation (1969).

The challenges discussed are particularly pronounced in municipalities located in the Arctic. Kantola and Tuulentie (2020, p. 5) note that Arctic cities are characterised by historically small populations. Urban growth in these areas is often due to the migration of residents, particularly young, educated women, from less densely populated regions to urban centres. This movement is compounded by international interests in Arctic tourism as well as by establishing military bases. Regarding the tourist sector, it should be noted that it is dominated not by family-owned businesses but rather by multinational corporations.[6] Additionally, the Arctic is a region coveted for its abundance of natural resources that influences, for example, the development of mines (Tolvanen *et al.* 2019, Bay-Larsen *et al.* 2018). Consequently, Arctic cities represent a complex intermingling of global, national, and local interests, presenting residents with the challenge of navigating a dense network of connections. In this vein, Brunet *et al.* (2014) observe that despite a marginal increase in local engagement within Arctic regions over the past 50 years, participation continues to vary significantly across disciplines, organisations, and geographical areas (refer to Kantola and Tuulentie 2020 for further discussion). In this context, there are, for example, visible tensions between Indigenous people and those who simply come to the North to live and work (Larsen 2018). Finally, it should be emphasised that low population density and the extent of urban areas—including both the very centre and districts located at great distances, as observed for example in Rovaniemi and Tromsø—are also not conducive to the effective organisation of participatory tools.

5.5 Participation Practices in the European Arctic

In this section, we present the results of analysis of research conducted in our case study cities. We examine various issues pertaining to participation, including the relationships between institutions and different actors that are critical for sustainable urban development in European Arctic cities, as well their attitudes towards sustainable development (SD). Furthermore, we discuss how the collaboration among state and non-state actors and participatory practices can be enhanced to align with SD policies more effectively. Our goal is to provide a comprehensive overview of participation—its strengths and weaknesses—across all cities studied, while also delving into the specifics of some case studies in greater detail.

Prior to commencing the analysis, it is imperative to acknowledge the observed transition from New Public Management to governance. This shift necessitates the incorporation of more progressive methods of participation, as well as the engagement of traditionally underrepresented groups such as the elderly, youth, immigrants, and Indigenous people. In this context, Bovaird, Löffler, and Diez define local governance as

> the set of formal and informal rules, structures and processes by which local stakeholders collectively solve their problems and meet societal needs. This process is inclusive because each local stakeholder brings important qualities, abilities and resources. In this process it is critical to build and maintain trust, commitment and a system of bargaining.
>
> (2003, p. 374)

Governance, therefore, is concerned with "institutional optimisation"—going beyond constitutional and legal frameworks (as outlined in Chapter 4) to encompass uncodified behavioural norms, including attitudes and beliefs. The determination of an optimal combination of structures, such as hierarchies and networks, is integral to the process of building institutions. While New Public Management focused heavily on inputs, outputs, and outcomes, governance emphasises the significance of ownership and participation in decision-making processes. Hence, the contemporary discourse on public governance reasserts the long-standing principle that the nature of processes is of substantial importance (Liddle 2021). As previously mentioned, governance is characterised by the involvement of multiple actors. It is not solely public agencies that exert influence on the formulation of rules and their execution; non-governmental actors, including businesses and media, also affect the quality of life in local communities and the general well-being of these communities through their interactions with other actors. At the same time in the European Arctic municipalities—where global, national, and local interest are crossed—the activity of some actors, for example, business and Indigenous peoples, is more intensive, on the other hand, and the focus on the land planning is more visible (Kantola, Tuulentie 2020) on the other hand.

All cities included in our case studies implement participatory mechanisms to varying extents. Among the seven cities considered, Rovaniemi and Tromsø are particularly noteworthy for their active and innovative approaches. This dynamism is attributed to at least two factors. Firstly, both cities are relatively large, enhancing their capacity for innovation. Secondly, they engage significantly on international platforms. This engagement is partly due to their roles as academic and tourist hubs and as sites for international institutions, primarily the Arctic Council Secretariat or events such as annual the Arctic Frontiers conference (see Chapter 6 for more details). Presently, both cities also boast a workforce comprising individuals from numerous countries. Specifically, Rovaniemi emphasises the involvement of youth and migrants (International Rovaniemi 2024a), while Tromsø, one of Norway's fastest-growing cities, focuses on citizen participation in transforming the urban landscape (Nyseth, Ringholm, and Agger 2019). According to Nyseth (2011) and Nyseth *et al.* (2010), Tromsø has a long-standing tradition of adopting experimental methods in urban planning. Since the 1990s, the city has introduced various collaborative planning projects to engage different actors and citizens in urban transformation initiatives. Notably, during 2015–2016, Tromsø implemented the

"Where is Tromsø going?" project to envision the city's future. This initiative was a component of a formal planning process associated with a municipal master plan. It involved diverse participatory activities, including public meetings, philosophical dialogues, city walks to showcase local heritage, cultural events, workshops, democratic cafés, seminars, and exhibitions. The project also featured films on urban planning themes shown at a cinema, accompanied by post-screening discussions, lectures, blog posts, and chronicles that were disseminated in local newspapers and through specially published "small papers" (Nyseth *et al.* 2019, p. 10). This does mean though that our other case studies do not have specific features related to participation and the inclusion of citizens in policy-making processes. In Kiruna (Sweden) and Kolari (Finland), many participatory activities, both implemented by authorities and initiated by citizens, were conducted in the shadow of the mining industry. In Luleå, despite relatively well-developed mechanisms, there has been a steady decrease in interest and participation (sometimes caused by organisational shortcomings, such as inconvenient meeting times during the day when people are working). In the cases of Akureyri and Nuuk, we observed that available participatory mechanisms are not perceived as efficient, and some of the interviewees indicated informal ways of communication as more important than formal interactions or procedures.

Firstly, the results of our analysis[7] indicate that the idea of sustainable urban development is, according to our interviewees, vague and complex. This may result both from the complexity of the term and the diverse legal regulations related to it (see Chapter 4), as well as from the fact that various actors approach the term differently. In this context, our interviewees indicated that "discussion about sustainable urban development is superficial" (researcher, Akureyri, 27.05.2022) and that the concept is understood differently by politicians and administrators (business representative, Tromsø, 25.04.2022). Moreover, "sustainability is often seen as a technical problem by governments, [a problem—KRM] for which there are technical solutions" (researcher, Kiruna, 17.01.2022). Against this background, citizens do not seem interested in implementing new ideas and solutions if they have to pay extra for them. This attitude impacts interest and participation in the issue. One administrator from Rovaniemi commented on this in the following way:

> Citizens want sustainability and sustainable city but only when they do not have to change own habits and behaviors. This is classical NIMBY [Not in my back yard –KRM] "I do that but if it makes my life more difficult, this not a good idea". So real participation still waits, it is yet to come influencing policy-making.
>
> (administration, Rovaniemi, 09.03.2022)

Secondly, the results of our analysis indicate a very complex constellation of actors engaged in urban development in the municipalities in question. As already pointed out, differing interests at the global, national, and local levels result in contradictory policies and approaches towards (sustainable) urban development. urban development. In this context, our interviewees pointed out the following matters. The interviewees indicated that different resources and capacities of actors influence their participation in and access to the process. This issue should be viewed, however, from several perspectives. First, the analysis of interviewees' statements indicates that the dominant groups influencing policy- and decision-making are generally still national/local politicians and officials. They not only have access to information but also to knowledge and resources that allow them to shape the participatory process. At the same time, many interviewees pointed out that national interests do not align with local ones, resulting in tensions and mutual accusations between the national government and local self-government sides. In some interviews, the issue of a lack of understanding of the problems of the North by politicians seated in parliaments located in the respective capital was raised (centre/periphery).

On the one hand, one of our interviewees from Greenland indicated that "Denmark still dominates most topics" (researcher, Greenland, 15.07.2022) and that "there is a mismatch between the resources that are local and important decisions that have regional or even global impacts" (researcher, Greenland, 29.04.2022). However, the results of our analysis also indicate that local politicians are divided. Some of them, educated in the capitals located in the South and having contacts there (or also holding the mandate of a councillor and a MP), not only have better knowledge about how public policies are created, but also a number of informal contacts that allow them to better influence the decision-making process. On the other hand, the results of our analysis show that the issues that particularly divided national, regional, and local politicians were those concerning the mining industry and the development of tourism. As one interviewee from Kolari, a municipality where there is an ongoing debate about reopening a mine, pointed out: "If you talk at the national level, people don't want the mine here" (politician, Kolari, 21.11.2022), while at the same time, in the municipality, there is a discussion about its development, as the Kolari's budget cannot be secured solely by income from tourism. The issue of tourism's strong impact on urban development was mainly raised by interviewees from Rovaniemi and Tromsø. In this context, they pointed to the increasing cost of living associated with the fact that many properties are allocated for tourist rentals, and services are tailored to their budgets and needs.

Secondly, within the context of resources and the North-South relations, respondents emphasised that it is much easier for cities such as Rovaniemi or Tromsø to present their case than it is for small municipalities such as Kolari. One of the interviewees addressed that issue in the following way:

> (…) a municipality the size of Kolari alone is not very much able to influence the development policy of the entire European Arctic region. Of course, for a small part, everyone together can influence development, but the measures of an individual municipality do not have a big impact.
>
> (politician, Kolari, 04.03.2022)

Thirdly, in one of the cities analysed—Kiruna—interviewees indicated that the dominant actor influencing policy- and decision-making is industry. This is especially visible in Kiruna's urban transformations, as the LKAB mining company is paying for the participatory processes accompanying the transformations. Our interviewees also indicated that there are joint group meetings organised every week in which civil servants from the town hall and representatives of LKAB take part.

Fourthly, the results of the conducted analysis indicate that "ordinary citizens" are among the weakest actors, both in the context of knowledge and available resources. Furthermore, many interviewees pointed out that participation in policy-making processes are organised in national languages, which excludes many people, including seasonal workers evident in many sectors of the current industry in the (European) Arctic.

The second issue raised by our interviewees in the context of relationships between institutions and different actors is of participation—both the attitude towards it and the organisation of the participatory process. In this context, our interviewees indicated several issues. On the one hand, in many interviews, especially those conducted with local politicians, the theme of "ordinary citizens" appeared. These citizens were deemed unprepared by many interviewed to make responsible decisions related to the SD of the municipality. In one of the interviews conducted in Kolari, residents were described as "stupid" and "loud" (politician, Kolari, 07.03.2022). The interviews also touched upon the issue of proper legitimisation of the decision-making process—significant from the perspective of the Nordic model assumptions. One of our interviewees from Tromsø indicated that too few people participate in the participatory activities organised by the city, and, therefore, decisions made in this way cannot be considered binding (politician, Tromsø, 20.03.2023). Additionally, our interviewees—particularly those from Finland—pointed to a characteristic culture of the North, which involves not talking about problems and difficulties. In this context, many people find it difficult to openly criticise authorities and officials during the meetings that the latter organise.

On the other hand, an extremely important theme raised by the respondents was the way the participatory process is organised. In this context, several aspects are important. First, the municipalities' economic circumstances must be taken into consideration when elaborating on the participatory processes they are implementing (or not). Several interviewees explained a situation in which many municipalities find themselves—a vicious circle. Smaller, remote municipalities in the North in particular are confronted with a comparably low population and, thus, with low tax revenues but an overall large territory to manage. Attracting a work force particularly in the public sector is difficult since they cannot pay the highest salaries in comparison to other municipalities and employers due to the described conditions. In terms of providing good public services and infrastructure, due to their limited human and financial capacities, some municipalities see themselves in competition with each other for private investment to cope with social and economic challenges. This situation often explains why debates are described as being dominated by industry and business representatives in contrast to different perspectives from local citizens, rights-holders, and other interests' groups (Garbis *et al.* 2023). Our interviewees shared that it is difficult to involve people because they often do not feel like their participation can make a difference. The participatory processes are not perceived as meaningful, which is why many citizens do not see how they could actually influence them. Hence, many citizens are reluctant to participate in the first place. This image is particularly strong for people who identify themselves as belonging to minorities, representing perspectives that might differ from the majority opinion, for instance, in the case of (Sámi) reindeer herders. Second, our interviewees indicated the complexity of issues related to SD. Also, many citizens have the feeling that if they ask critical questions in a participatory process, they will be accused of being "against" sustainability and green transformation. One researcher from Kiruna noted:

> For the citizens, it becomes harder to feel that they can actually influence processes—not very democratic processes. It is also hard to go more into detail, talk about complexity. There is a risk of raising their perspectives.
>
> (Researcher, Kiruna, 20.12.2022)

Additionally, the analysis of the research material collected in Finland and Iceland suggests that the political culture slowly changes; some of the most important decisions are still made behind closed doors—in saunas or at swimming pools. One of our interviewees commented on this social phenomenon: "There is in Iceland a tradition that you always go to the swimming pool to sit in the hot pot. Local politics are discussed [there] and they are discussed in length" (Citizen, Akureyri, 13.07.2022).

Furthermore, the analysis of interview content indicates that there are difficulties in organising the participatory process. Also in this context, at least two themes are significant. On the one hand, it should be emphasised that citizen participation in decision-making processes related to spatial planning is mandatory in Nordic countries (see more on the legal requirements in Chapter 4). At the same time, our interviewees pointed out that officials often fulfil the mandatory minimum, without considering whether planning "public hearings" in the morning hours, when many citizens are at work or school, is appropriate. This leads to a situation where the same individuals, the "usual suspects", repeatedly have a voice in local matters (Boulianne 2018).

On the other hand, there is a question of who designs and leads the participatory agendas and processes. For example, in the case of urban transformations in Kiruna, the dominating role—as already indicated—is performed by the state-owned mining company which is paying for the processes. Many of our interviewees described these particular circumstances for participation to be hindering factors for meaningful citizen involvement. In terms of more innovative participatory tools, the company is often hesitant to implement them as they are linked to higher costs. Discussions are only facilitated at a superficial level that do not allow for in-depths exchanges. Overall, sustainability tends to be described as a technical problem, for which the industry could cope with technical solutions. In this regard, many interviewees shared that they see a strong focus on the economic dimension and less on social and environmental questions of sustainable urban development. What is more, our cases illustrate how global and national agendas unfold at the local level. We identify a gap between the governance levels and within them regarding knowledge exchange. Decisions and commitments are often made at higher governance levels, and the implementation takes place at the local level, where the tools to cope with these tasks are often limited or even entirely missing. These impressions can be further linked to a more critical note on the municipal autonomy for spatial planning. Referring again to the Kiruna case, it is not only a positive characteristic providing the municipality with certain freedoms, but, due to the many tasks that need to be fulfilled by the municipality linked to the urban transformation which is needed to continue mining (being a national interest), the municipality sees itself confronted with many, often overwhelming tasks. Our interviewees indicated that many issues, such as migration policy, are still treated as indirectly related to SD (researcher, Tromsø, 08.03.2023).

5.6 Conclusions

Based on our findings, we can draw several conclusions. First, analysis of the collected data points to a gap between the assumptions of the Nordic model and the governance practices in Arctic municipalities. Although the model formally presupposes trust, inclusion, equality, and transparency in decision-making processes, our research findings reveal an imbalance among actors. This is primarily due to the knowledge(s) and resources they possess. Consequently, policy- and decision-making processes continue to be dominated by politicians, administrators, and industry representatives, which was particularly evident in the case of Kiruna. Furthermore, our findings highlight a gap between the formal and informal rules of governance in European Arctic municipalities. Despite the enforcement of formal rules and laws, many decisions are made behind closed doors as the result of informal conversations and contacts.

Moreover, our analysis shows that the distinction between the governing and the governed blurs. Viewing governance as a complex process in which public authorities and private actors interact in pursuit of agreed goals, it becomes more challenging to define some actors as governing and others as governed. Instead, the pertinent question is how, when and where different actors are expected to actively contribute to the governance of society (Torfing *et al.* 2012, p. 151). Furthermore, from the perspective of the governance paradigm, in European Arctic municipalities, the distinction between public and private actors becomes less relevant as a reference point for determining whether an actor should be considered a legitimate participant in a governance process. The legitimacy of various actors has less to do with their sectoral affiliation than with the degree to which they are affected by governance processes or possess knowledge and resources relevant to contributing to the definition or resolution of governance tasks.

Additionally, the results of our analysis indicate that, despite legal regulations and guidelines (see Chapter 4), the concept of sustainable urban development remains vague and complex. This chapter's analysis of local governance implementation confirms that local actors perceive it in various ways, often regarding it as an empty signifier. This ambiguity affects participatory processes, which tend to be neither inclusive nor innovative. Instead, these processes are often structured to maintain the existing status quo.

Finally, the results of our study confirm that governance in cities located in the European Arctic challenges the well-established distinction between political and administrative tasks and the notion that the former precede the latter. By acknowledging that all phases of a governance process involve both political and administrative tasks, it becomes much more complex to distinguish between the types of norm-based actions to be performed by politicians and public administrators, respectively. Although it is still somewhat valid to consider politicians as policy-makers and public administrators as policy implementers, it is increasingly difficult to establish a clear division of labour between the two. Overall, our findings underscore the complexity of governance dynamics in European Arctic cities, highlighting the need for a nuanced understanding of the roles that actors take, policy- and decision-making processes and the interplay between formal and informal governance structures.

Acknowledgements

This chapter was written by Katarzyna Radzik-Maruszak. The chapter was reviewed by Jacqueline Götze, Michał Łuszczuk, Arne Riedel, Dorothea Wehrmann and an external expert. All authors of this book read and accepted the content of this chapter.

Notes

1 One of our cases is the city of Nuuk, which is located in Greenland. In this chapter, we also mention Denmark as this country is a part of the Nordic Model.

2 This is important because, in the Nordic countries, professional training and preparation play a crucial role in decision-making and the execution of specific functions. In this context, the knowledge and competencies of citizens are frequently called into question, impacting participatory processes. For further details, refer to the subsequent discussion.

3 On the one hand, there are only three big cities Tampere, Helsinki and Turku, and four smaller municipalities Kärkölä, Pirkkala, Tuusula, and Puolanka with governing mayors. On the other hand, it should be noticed that more than 20% of Finns live in these seven units.

4 With an exception of Iceland.

5 These are critical in the process of spatial planning. See *The spatial planning systems in the Nordic region:* https://archive.nordregio.se/Metameny/About-Nordregio/Nordic-working-groups/nwgcityregions/The-spatial-planning-systems-in-the-Nordic-region/index.html [Accessed 23 May 24].

6 However, it depends on the country. For example, in Finland small family companies tend to dominate (see Komppula 2013).

7 See Chapter 1 for more information on the methodology and interviewees.

References

Akureyri, 2022. Interview with researcher from Akureyri, interview conducted 27.05.2022.

Akureyri Municipality 2024a. Available from: https://www.akureyri.is/en/administration/about-akureyri/town-council [Accessed 20.08.24].

Akureyri Municipality 2024b. Available from: https://www.akureyri.is/is/stjornkerfi/akureyri/vinabaeir [Accessed 20.08.24].

Andreasson, U., 2017. *Trust–the Nordic gold.* Nordic Council of Ministers.

Antikainen, A., 2006. In search of the Nordic model in education. *Scandinavian Journal of Educational Research*, 50 (3), 229–243.

Anttiroiko, A.-V., and Haveri, A., 2022. Local government and the COVID-19 pandemic. *In:* Silva, C. N., ed., *Local government and the COVID-19 Pandemic*. Cham: Springer, 59–83.

Arnstein, S. R., 1969. A ladder of citizen participation. *Journal of the American Institute of Planners*, 35 (4), 216–224.

Baldersheim, H., Houlberg, K., Lidström, V., and Hlynsdottir, E. M., 2019. *Local autonomy in the Nordic countries: A report for the Norwegian Association of Local and Regional Authorities*. Kristiansand: Universitetet i Agder.

Bay-Larsen, I., Skorstad, B., and Dale, B., 2018. Mining and Arctic communities. *In:* Dale, B., Bay-Larsen, I., Skorstad, B., eds., *The will to drill – Mining in Arctic communites*. Cham: Springer Polar Sciences. Available from: https://doi.org/10.1007/978-3-319-62610-9_1.

Beierle, T., and Cayford, J., 2002. *Democracy in practice: Public participation in environmental decisions*. New York: Routledge.

Bherer, L., Dufour, P., & Montambeault, F., 2023. What is informal participation? Introduction to the special issue: Quietly standing out: Understanding informal forms of political engagement. *International Journal of Politics, Culture, and Society*, 36 (1), 1–16.

Blom-Hansen, J., & Heeager, A. (2010). Denmark: Between local democracy and implementing agency of the welfare state. In: Frank Hendriks, Anders Lidström John Loughlin, ed. *The Oxford Handbook of Local and Regional Democracy in Europe*. Oxford University Press, Oxford.

Blossing, U., Imsen, G., and Moos, L., 2014. *The Nordic education model*. Dordrecht: Springer Netherlands.

Boulianne, S., 2018. Beyond the usual suspects: Representation in Alberta climate dialogue's deliberative exercises. *In:* Hanson, L., ed., *Public deliberation on climate change: Lessons from Alberta Climate Dialogue*. Edmonton: Athabasca University Press.

Bovaird, T., Löffler, E., and Parrado-Díez, S., 2003. Emerging practices in network management at local levels in Europe. *In:* Salminen, A., ed., *Governing networks. EGPA yearbook*. Amsterdam: International Institute of Administrative Sciences Monographs, 371–386.

Brunet, N. D., Hickey, G. M., and Humphries, M. M., 2014. The evolution of local participation and the mode of knowledge production in Arctic research. *Ecology and Society*, 19 (2), 69. https://doi.org/10. 5751/ES-06641–190269.

City of Rovaniemi Administrative Regulations 2024. Available from: https://www.rovaniemi.fi/loader.aspx?id=f51cc0ff-fbba-4173-ba03-510ad4ae3f39 [Accessed 20.08.24].

Denmark. Unitary Country 2023. Available from: https://www.sng-wofi.org/country-profiles/denmark.html [Accessed 20.08.24].

Dodds, K., and Woodward, J., 2021. *The Arctic: A very short introduction*. Available from: https://doi.org/10.1093/actrade/9780198819288.003.0006 [Accessed 20.08.24].

Edvardsson, I. R., 2014. A small university and knowledge-based development: A case of Northern Iceland. *International Journal of Knowledge-Based Development, Inderscience Enterprises Ltd*, 5 (2), 131–151.

Esping-Andersen, G., 1990. *The three worlds of welfare capitalism*. Polity.

EU's Arktiskontor i Grønland 2024. Available from: https://denmark.representation.ec.europa.eu/eus-arktiskontor-i-gronland_da [Accessed 20.08.24].

Eythórsson, G. T., Gløersen, E., and Karlsson, V., 2015. Municipalities in the Arctic in challenging times. West Nordic local politicians and administrators on municipal structure, local democracy, service provision and adaptive capacity in their municipalities. Results from a survey among elected local officials and bureaucrats in the Faroe Islands. Greenland and Iceland: A project supported by the Arctic Co-operation Programme 2012–2014, University of Akureyri, Spatial Foresight GmbH, University of Akureyri Research Centre & West Iceland Regional Office.

Eythórsson, G. T., Kettunen, P., Klausen, J. E., and Sandberg, S., 2018. Reasons for inter-municipal cooperation: A comparative analysis of Finland, Iceland and Norway. *In:* Swianiewicz, P., Teles, F., eds., *Inter-municipal cooperation in Europe: Institutions and governance*. New York: Palgrave Macmillan, 105–129.

Feeley, M., and Langford, M., 2022. Nordic exceptionalism and the legal complex. *In:* Feeley, M., and Langford, M., eds., *The limits of the legal complex*. Oxford: Oxford University Press, 13–67.

Fukuyama, F., 2014. *Political order and political decay*. New York: Macmillan.

Gad, U. P., 2022. Nuuk as a diplomatic scene. Protocol, pitfalls, and practice, DIIS POLICY BRIEF. Available from: https://www.diis.dk/en/research/nuuk-as-a-diplomatic-scene [Accessed 20.08.24].

Garbis, Z., *et al.*, 2023. Governing the green economy in the Arctic. *Climatic Change* 176, 33. https://doi.org/10.1007/s10584-023-03506-3 [Accessed 20.08.24].

Goldsmith, M., and Larsen, H., 2004. Local political leadership: Nordic style. *International Journal of Urban and Regional Research*, 28 (1), 121–133.

Greve, C., Lægreid, P., and Rykkja, L. H., 2016. Introduction. The Nordic model in transition. *In:* Greve, C., Lægreid, P., and Rykkja, L. H., eds., *Nordic administrative reforms: Lessons for public management*. London: Palgrave Macmillan, 1–21.

Hall, P., Kettunen, P., Löfgren, K., and Ringholm, T., 2009. Is there a Nordic approach to questions of democracy in studies of network governance? *Local Government Studies*, 35 (5), 515–538.

Hansen, K. G., 2015. Greenland is rethinking the 2009 merging of municipalities. *Nordregio News*, (3), 14. https://norden.diva-portal.org/smash/get/diva2:860942/FULLTEXT01.pdf [Accessed 19/04/24].

Haveri, A., 2015. Nordic local government: A success story, but will it last? *International Journal of Public Sector Management*, 28 (2), 136–149.

Hesseln, H., Silvén, P., and Siemińska, K., 2013. UArctic: Evaluating 10 years of collaboration. *The Polar Journal*, 3, 204–226. Available from: https://doi.org/10.1080/2154896X.2013.776295

International Rovaniemi 2024a. Available from: https://international.rovaniemi.fi/en

International Rovaniemi 2024b. Available from: https://international.rovaniemi.fi/en/Services/About-Us/Twin-Cities

Johansson, H., 2022. Urban and local social policies in the Nordic countries. *In:* Kazepov, Y., Barberis, E., Cucca, R., and Mocca, E., eds., *Handbook on urban social policies: International perspectives on multilevel governance and local welfare*. Northampton, MA: Edward Elgar Publishing, 414–428.

Kolari, 2022. Interview with politician from Kolari, interview conducted 21.11.2022.

Kantola, S., and Tuulentie, S., 2020. Participation in a large Arctic city – the possibilities of PPGIS for improving interaction. *Polar Geography* 43 (4). DOI: 10.1080/1088937X.2020.1767709

Karlsson, V., and Eythórsson, G. T., 2022. Municipal amalgamations and local housing prices. *Region*, 9 (1), 101–114. DOI: 10.18335/region.v9i1.360

Kiruna Municipality 2024. Available from: https://kiruna.se/kommun--demokrati/kommunens-organisation.html [Accessed 20.08.24].

Koivurova, T., and Hasanat, M. W., 2009. The climate policy of the Arctic Council. *In:* Koivurova, T., Keskitalo, E., Bankes, N., eds., *Climate governance in the Arctic*. Springer: Dordrecht, 51–75. https://doi.org/10.1007/978-1-4020-9542-9_3

Komppula, R 2013. Success and growth in rural tourism micro-businesses in Finland: Financial or life-style objectives?. *In:* Rhodri Thomas, ed., *Small firms in tourism.* London: Routledge, 115–138.

Lapin Liito 2024. Available from: https://www.lapinliitto.fi/en/information/general-information-about-lapland/ [Accessed 20.08.24].

Lehtonen, P., and Radzik-Maruszak, K., 2024. Inclusion as ownership in participatory budgeting: Facilitators' interpretations of public engagement of children and youth. *Critical Policy Studies*, 18 (1), 73–91. https://doi.org/10.1080/19460171.2023.2192412

Leinonen, J., 2012. *Monelta suunnalta on suitsia suussa, mutta niiden kanssa on elettävä´ – Johtamisen liikkumavara kunnanjohtajan silmin (Managerial Latitude in Finnish Local Government: Municipal Managers' Perspective*), Acta 232. Helsinki: Association of Finnish Local and Regional Authorities.

Larsen, R. K. 2018. Impact assessment and indigenous self-determination: A scalar framework of participation options. *Impact Assessment and Project Appraisal*, 36 (3), 208–219. DOI: 10.1080/14615517.2017.1390874

Liddle, J., 2021. New public governance. *In:* List, R. A., Anheier, H. K., Toepler, S., eds., *International encyclopedia of civil society*. Cham: Springer, 1–6. Available from: https://link.springer.com/referenceworkentry/10.1007/978-3-319-99675-2_9580-1

Löfgren, K., and Ringholm, T., 2009. Introduction: New network modes of Nordic local governance. *Local Government Studies*, 35 (5), 505–514.

Manzi, T., Lucas, K., Jones, T. L., and Allen, J., 2010. Understanding social sustainability: Key concepts and developments in theory and practice. *In:* Manzi, T., Lucas, K., Jones, T. L., and Allen, J., eds., *Social sustainability in urban areas: Communities, connectivity and the urban fabric*. London: Earthscan, 1–28.

Martela, F., Greve, B., Rothstein, B., and Saari, J., 2020. The Nordic exceptionalism: What explains why the Nordic countries are constantly among the happiest in the world. *In:* Helliwell, J. F., Layard, R., Sachs, J. D., and De Neve, J. E., eds., *World happiness report 2020. Sustainable developments solutions network.* Sustainable Development Solutions Network, 129–146. Available from: https://worldhappiness.report/ed/2020/#rea [Accessed 20.08.24].

Mouritzen, P. E., and Svara, J. H., 2002. *Leadership at the apex: Politicians and administrators in Western local governments.* Pittsburgh: University of Pittsburgh Press.

Nilsen, T., 2022. Murmansk terminates sister city relations with Akureyri, Available from: https://thebarentsobserver.com/en/life-and-public/2022/12/murmansk-terminates-sister-city-relations-akureyri [Accessed: 1.07.2024].

Nilson, H. R., 1997. *Arctic environmental protection strategy (AEPS): Process and organization, 1991–1997. An assessment.* OSLO: NORSK POLARINSTITUTT, No. 103.

Norwegian Local Governments Sector Outlook 2023. Available from: https://www.scopegroup.com/dam/jcr:ff37f1de-3531-4de3-9624-f9bc97d605af/Scope%20Ratings_Norwegian%20local%20governments_Sector%20Outlook%202023_Final.pdf [Accessed 20.08.24]

Nyholm I., 2006, Developing local democracy – the state and challenges of Finnish local democracy in the postmodern era. *Finnish Local Government Studies*, 4, 340–352.

Nyholm I., Haveri A., 2009, Between government and governance. Local Solutions for reconciling representative government and network governance. *Local Government Studies*, 35 (1), 109–124. DOI: 10.1080/03003930802574516.
Nyseth, T., 2011. The Tromsø experiment: Opening up for the unknown. *Town Planning Review*, 82 (5), 573–593.
Nyseth, T., Ringholm, T., and Agger, A., 2019. Innovative forms of citizen participation at the fringe of the formal planning system. *Urban Planning*, 4 (1), 7–18. https://doi.org/10.17645/up.v4i1.1680
Pawłowska, A., *et al.*, 2021. *Social councils and committees as (not-quite) present actors in local decision-making processes*. Warsaw: Wydawnictwo Naukowe Scholar.
Radzik-Maruszak, K., 2019. *Rada gminy jako uczestnik lokalnego współrządzenia. Przykład Anglii, Finlandii, Polski i Słowenii.* Warszawa: Wydawnictwo Naukowe Scholar.
Record Population Increase in Iceland 2023. Available from: https://www.icelandreview.com/society/record-population-increase-in-iceland/ [Accessed 17.09.23].
Record Population Increase in Iceland 2023. Available from: https://www.icelandreview.com/society/record-population-increase-in-iceland/ [Accessed 20.08.24].
Rovaniemi City Board 2024. Available from: https://www.rovaniemi.fi/Kaupunki-ja-paatoksenteko/Paatoksenteko/Kaupunginhallitus [Accessed 20.08.24].
Schrage, J., & Kjærås, K., 2022. How do cities challenge patterns of demand? Characterising the local governance of climate change in Nordic cities. *Environment and Planning C: Politics and Space*, 40 (7), pp. 1473–149.
Smith, G., 2009. *Democratic innovations*. Cambridge: Cambridge University Press.
Sørensen, E. (2022). Democratic network governance. *In*: Ch. Ansell, J. Torfing, ed. *Handbook on Theories of Governance*. Edward Elgar Publishing, Cheltenham, UK, Northampton MA, USA, 462–471.
Stende, T., 2017. *Is the Nordic Region best in the world?* Analysis No. 02.17, Nordic Council of Ministers.
Stoker, G., 2011. Was local governance such a good idea? A global comparative perspective. *Public Administration*, 89 (1), 15–31.
Svels, K., and Thuesen, A. A., 2024. Input legitimacy of bottom-up fishery governance: Lessons from community-led local development in two Nordic EU countries. *Sociologia Ruralis* 64, 445–469, https://doi.org/10.1111/soru.12479
The Barents Observer 2022. The capital of Lapland freezes partnership with Murmansk. Available from: https://thebarentsobserver.com/en/life-and-public/2022/03/capital-lapland-freezes-partnership-murmansk [Accessed 20.08.24].
The spatial planning systems in the Nordic region 2024. Available from: https://archive.nordregio.se/Metameny/About-Nordregio/Nordic-working-groups/nwgcityregions/The-spatial-planning-systems-in-the-Nordic-region/index.html [Accessed 20.08.24].
Tolvanen, A., *et al.*, 2019. Mining in the Arctic environment – A review from ecological, socio-economic and legal perspectives. *Journal of Environmental Management*, 233, 832–844.
Torfing, J., Peters G.B, Pierre J., Sørensen E., 2012. *Interactive governance: Advancing the paradigm*. Oxford University Press, New York.
Tornio 2024. Available from: https://www.tornio.fi/en/city-of-tornio/information-on-tornio/international-cooperation-and-sister-cities/ [Accessed 20.08.24].
Tromsø City Council 2024. Available from: https://tromso.kommune.no/politikk/politisk-organisering/kommunestyret [Accessed 20.08.24].
Tromsø Executive Commission 2024. Available from: https://tromso.kommune.no/politikk/politisk-organisering/formannskapet-og-hovedutvalgene [Accessed 20.08.24].

Trydegård, G., and Thorslund, M., 2010. One uniform welfare state or a multitude of welfare municipalities? The evolution of local variation in Swedish elder care. *Social Policy & Administration*, 44 (4), 495–511.

Tunström, M., 2019. Urban social sustainability policies in the Nordic region: A repackaging of the welfare state model? *In:* Reza Shirazi, M., and Keivani, R., eds., *Urban social sustainability: Theory, policy and practice*. New York: Routledge, 42–58.

Vakkala, H., Sinervo, L. M., and Jäntti, A., 2021. Local self-government in Finland. *In:* Brezovnik, B., Hoffman, I., Kostrubiec, J., eds., *Local self-government in Europe*. Maribor: Institute for Local Self-Government, 173–205.

Wennemo, I., 2014. *Det gemensamma: om den svenska välfärdsmodellen*. Stockholm: Premiss förlag.

Wetterberg, G., 2004. *Den kommunala självstyrelsen*. SNS Förlag: Avesta.

White, C., 2000. Depoliticising development: The uses and abuses of participation. *In:* Eade, D., and Pearce, J., eds., *Development, NGOs, and civil society*. London: Oxfam GB, 142–155.

6 Cooperation between Cities

6.1 Introduction

How and to what extent can (transnational) cooperation with and between cities in the European Arctic contribute to sustainable urban development in line with local needs and global policies? In this chapter, we analyse the potentials and challenges for cooperation specifically in view of the cities that we investigated in our case studies. We explore the extent to which cooperation in the Arctic regions

1 may stimulate and enhance bringing together different knowledges and perspectives in the context of sustainable urban development and
2 allows for the better alignment of sustainable development policies designed at various governance levels.

In this context, our primary effort is to contribute to a better understanding of how the pursuit of the current global goals and strategies of sustainable development can be advanced in remote regions.

In what follows, we first introduce the scholarly debate on (transnational) cooperation with and among non-state actors in the Arctic. Second, we present an empirical discussion of the theoretical assumptions on (transnational) cooperation introduced in Chapter 2. Thereby, we also refer to the insights from Chapter 3, endeavouring to track the implementation of global governance instruments, and to Chapter 4, which explores how the Nordic model of governance shapes and influences urban participation in cooperation at the local level and across governance levels and regions, as well as the structure and nature of these collaborative efforts.[1]

In the first part of this chapter, we introduce the current state of (transnational) cooperation in the Arctic by considering, in particular, the urban dimension. We first provide a general overview of the relevance ascribed to (transnational) cooperation in the Arctic since the end of the Cold War, regional approaches to the Arctic and the most visible actors in (transnational) cooperation. Second, we address a gap in research on cities by focusing specifically on the challenges and potentials of (transnational) cooperation between remote European Arctic cities, thus considering the theoretical considerations introduced in Chapter 2.

DOI: 10.4324/9781003284864-6

In the second part of this chapter, we present the results of our qualitative analysis and discuss the insights shared in the 80 semi-structured interviews from our seven case study cities.[1] This chapter presents three main findings. Firstly, at the local level (transnational), cooperation in the European Arctic has been significantly less intense and extensive than we had expected, suggesting that it does not significantly enhance the alignment of sustainable development policies designed at various governance levels. This is especially relevant in the context of environmental, economic, and social sustainability of urban areas, where cohesive and integrated policies are crucial (Dempsey *et al.* 2011, Jacquier 2005). The lack of significant city-to-city and transnational cooperation therefore implies a missed opportunity for leveraging shared knowledges and some resources and for addressing common challenges in a more harmonised/aligned manner. Moreover, we observe that transnational cooperation in the Arctic seems so far to be rather pushed by actors engaged in single local initiatives and not orchestrated by national or regional authorities.

Secondly, the potential of transnational cooperation to stimulate and enhance the integration of different knowledges and perspectives in the context of sustainable urban development in remote regions also proved to be rather limited. Despite common or similar challenges in urban development, it appears that cooperation for the exchange of best practices, fostering synergies from diverse perspectives or jointly advocating for specific solutions in international institutions is not being developed nor recognised by citizens. The research results indicate that gradually developed forms of regional (transnational) cooperation, primarily the Arctic Mayors Forum (AMF), are currently more engaged in grappling with external crises and the limitations of their potential, rather than serving as support for more effective local implementation of global frameworks related to climate change and sustainable development. This signals a misalignment between the goals of the existing transnational initiatives and the broader goals of sustainable development.

Thirdly, our analysis reveals the significant impact of external factors on transnational cooperation, notably the COVID-19 pandemic (2020–2023) and the international crisis due to Russia's full-scale invasion of Ukraine (February 2022). These factors are crucial to better understand the observed limitations in (transnational) cooperation, as they have reshaped the dynamics of collaboration in the Arctic at large and disrupted existing or developing collaboration channels between cities, such as the AMF and other partnerships. Furthermore, external factors have caused states' national governments to (re)dominate the Arctic arenas due to their focus on traditional security issues and have also shifted regional priorities and perceptions of challenges, further complicating the landscape of cooperation in the Arctic.

All findings are instrumental in shaping our understanding of the dynamics of (transnational) cooperation in the Arctic region, highlighting both its strengths and its limitations. Furthermore, they offer a more detailed perspective on the interplay between local initiatives and broader governance structures, providing valuable insights into how various actors with interests in the Arctic interact with and within this unique geopolitical context. The conclusions presented at the end of this chapter are not only crucial for the overall argument of this book but also serve as a foundational empirical basis for developing potential pathways for sustainable urban development in the Arctic in the next chapter (Chapter 7).

6.2 (Transnational) Cooperation in the (European) Arctic

Many cities in the Arctic differ from the dominant idea of an urban environment, which is often shaped by cities, such as the capitals in the South (Nyseth 2017). For instance, Arctic cities and urban centres are smaller and more remote but fulfil the same functions and provide similar services than their counterparts in other world regions (see Chapter 1). Systematic investigations and approaches on how specifically local perspectives and priorities can be included in policy-making across all levels in remote regions are still missing and urgently needed to advance policy alignment. This Arctic perspective is necessary to capture global sustainable urban development in its diversity. In this book, we thus address a gap in current research on cooperation among cities (such as Bouteligier 2012, Curtis 2014 and 2016, Gordon and Johnson 2018, Pipa and Bouchet 2020), which has focused mainly on the question of how large city networks can combine forces to enhance their influence in global decision-making. In this section, we first discuss how transnational cooperation has been evolving in the Arctic in the last decades and will then relate this to municipal networks and to the challenges and potentials of (transnational) cooperation between Arctic cities.

6.2.1 Cooperation in the Arctic

For Indigenous peoples, the Arctic has always been a transnational space. They have lived in the Arctic regions for thousands of years and collaborated across national borders before and after those were established. With the end of the Cold War, also the Arctic-rim states considered the Arctic in political terms "as a common space" (Knecht and Keil 2013, p. 22). The Arctic became "one of the world's most stable and cooperative regions" (Zellen 2020), and peaceful (transnational) cooperation across the circumpolar north was often labelled as "Arctic exceptionalism" (Lackenbauer and Dean 2020). This understanding of collaborative governance was also reflected in the framing of the Arctic as "One Arctic" (U.S. Arctic Council chairmanship program 2015–2017). In the 2010s and following the sentiment that many Arctic concerns are of "transnational nature" (Wilson Rowe 2018, p. 124), the Arctic region was increasingly addressed through regional approaches—despite the geopolitical relevance of the region and all regional differences.[2] Instead, stereotypical conflict narratives were "more likely to be peddled by and for actors outside the region rather than in it" (Kuus 2023, p. 352).

Transnational cooperation has also been essential to the regional approach of the Arctic Council, the main intergovernmental forum in the Arctic (at least until 2022). When the Arctic Council was established in 1996, Indigenous Peoples Organisations (IPOs) were given a special say, and six IPOs have the status of permanent participants (Wilson 2019). Over time, a growing number of other non-state actors with observer status have also been able to shape the work of the Arctic Council, especially in its working groups and task forces (Wehrmann 2017). The maintenance of peace, its "landmark" scientific publications (The Northern Forum 2015) and the negotiation of three legally binding agreements of regional scope[3] under its auspices are seen as the main achievements of the Arctic Council. Particularly due to the unique acknowledgement of Indigenous peoples (Gamble and Shadian 2017, Knecht 2013), however, the Arctic Council has often been hailed as a success case for transnational cooperation (Wehrmann 2020). After Russia's illegal annexation of Crimea in 2014, "civilian academic and civil society knowledge-based cooperation [remained] quite resistant" (Bertelsen 2019, p. 254) and there was widespread agreement to continue cooperation with Russia in the Arctic Council (Käpylä and Mikkola 2015). This changed with Russia's full-scale invasion of Ukraine in February 2022 and the decision of seven members of the Arctic Council to pause cooperation with Russia (Łuszczuk *et al.* 2023).

In addition to the Arctic Council, transnational cooperation has also a long tradition in Arctic ministerial and parliamentary groupings, such as the Nordic Council of Ministers (NCM) and the Standing Committee of the Parliamentarians of the Arctic Region (SCPAR) (Łuszczuk 2015). However, transnational knowledge-based relations are in particular considered to be "the most resilien[t] in light of political crisis and conflict" (Bertelsen 2019, p. 260). It is in this context, specifically with regard to knowledge-based cooperation, that research institutions, such as the University of the Arctic (UArctic) network[4] and the International Arctic Science Committee (IASC),[5] as well as the Arctic Economic Council (AEC)[6] as a business alliance, are most visible in the Arctic. Also, non-Arctic actors, such as the European Union, play an important role for cross-border cooperation in the European Arctic as its funding programmes and frameworks facilitate(d) cooperation with Russian and North American partners as well (Stępień and Koivurova 2017).[7]

The Arctic "scene of much transnational scientific and knowledge-based collaboration between public, private, and civil society actors" (Bertelsen 2019, p. 253) is particularly visible at Arctic conferences (Steinveg *et al.* 2024), which "play a large role in Arctic Affairs" as they promote "mutual understanding and respect", which is essential for Arctic cooperation (Sellheim and Menezes 2022, p. 5). In this way, the Arctic Circle Assembly in Reykjavik is considered the "key international gathering of Arctic expertise" (Kuus 2023, p. 354). The Arctic Circle was formed by societal, academic, and business institutions in 2013 on the initiative of the former Icelandic president, Ólafur Ragnar Grimsson, as a forum to "facilitate dialogue and build relationships to address rapid changes in the Arctic" (Knecht 2013, p. 176). The Arctic Frontiers conference in Tromsø is the other most prominent annual Arctic conference, which was established in 2006 "when it organised the first global conference on economic, societal and environmental sustainable growth in the North" (Sellheim and Menezes 2022, p. 69). During the COVID-19 pandemic, however, (transnational) cooperation in the Arctic weakened. While most actors "knew each other already and could rely on the trust built up before the pandemic", due to travel restrictions, most Arctic meetings took place in virtual format, which removed "the element of humanity from interpersonal communication" (Kuus 2023, p. 353).

Aside of these institutionalised formats facilitating transnational exchange and cooperation in the Arctic, (informal) professional networks in particular are described as a "milieu of [...] trust" (Kuus 2023, p. 354) in the Arctic. Arctic networks are highly international, often specialised and most individuals involved have known each other for years. With the Arctic regions located so far away from capital cities, the scientists, diplomats, civil servants, businesspeople, and civil society actors who are usually involved in these Arctic networks due to their professional expertise do not necessarily live in the Arctic. For so-called "Arcticians", the specific intellectuals of circumpolar governance, their social expertise, their ability to "synthesize claims from multiple fields" and "to wear multiple hats" in order to communicate across disciplinary and national contexts have been considered crucial (Kuus 2023, p. 355). Such networks expand rapidly and informality is considered an integral facet of Arctic networks", which is built on the "social trust cultivated over years of continuous interaction" (Kuus 2023, p. 349).

6.2.2 Cooperation between Arctic Cities

As we have shown above, despite its remoteness, the Arctic is often seen as a "hub of transnational connections" (Palosaari 2012, p. 16), with interactions between state and non-state actors taking place at different sites ranging from transnational IPOs and regional forums such as the Arctic Council and the Barents-Euro-Arctic Council to international conferences and networks between Arctic cities. However, imaginations of the Arctic often do not include Arctic urban places, and similarly, Arctic urbanism is an understudied topic in urban and Arctic studies (Haapala 2022, Kenny 2017, Nyseth 2017). For example, one of the first academic books exploring urban sustainability in the Arctic was published in 2017 (Laruelle and Orttung 2017). Nevertheless, the Arctic, like other regions, is subject to the global trend towards urbanisation (Kantola and Tuulentie 2020), with the majority of the Arctic population already living in urban environments (Orttung and Suter 2020). Different to the "historical waves of Arctic urbanisation"[8] (Laruelle 2019), however, today climate change in particular is often seen as shaping the urban future of Arctic and sub-Arctic cities, making "the inclusion of local needs and interests in broader discussions on climate security and resilient cities" even more important (Filimonova 2023, p. 3).

In this way, the resilience of Arctic cities, their mitigation and adaptation strategies are also key concerns for local authorities (Filimonova 2023, Kenny 2017). Particularly with regard to adaptation, more "comprehensive planning agendas for the Arctic that balance resilient and sustainable development with the presented by climate change" (Kenny 2017, p. 143) and "access to and availability of technology and scientific information are essential determinants of municipal capacities to develop and implement climate adaptation policies" (Filimonova 2023, p. 3). However, the different definitions of cities and the diversity of urban places in the Arctic make it difficult to compare policies and planning agendas.[9] It is, therefore, challenging to measure Arctic cities by the same standard (Orttung and Suter 2020). Nevertheless, some studies have identified characteristics that Arctic cities have in common, such as “the large range of settlement types, but also the unique historical and socio-economic conditions” (DiNapoli and Juul 2020, p. 13). Others stressed the harsh environment, which “limits access, makes energy expensive, requires intensive attention to infrastructure, reduces the availability of labour, mandates reliance on vulnerable food supply chains, and threatens the maintenance of a strong community culture”, and described “the development of indigenous societies, resource extraction, and protecting national security” as main drivers for development in the Arctic (Orttung and Suter 2020, p. 14). In contrast, others emphasise the individuality of (Arctic) cities and consider the development of urban places as “always temporary, localised and dynamic” (Haapala 2024, p. 2).

Despite the differences, the intensification of sub-national collaboration in the Arctic regions, with the Arctic Mayors’ Forum and the Arctic Urban Regional Cooperation Programme as prominent examples, however, corresponds to the growing number of transnational municipal networks (TMNs) that focus on environmental sustainability beyond the Arctic.[10] More generally, municipal networks can be perceived as “a form of bottom-up governance” by local governments (Dumała *et al.* 2021, p. 2), with varying degrees of institutionalisation and formalisation. While the activities of the cities engaged in TMNs are restricted amongst others by national regulations and public funding, they can still be considered non-state actors “because their activities are not driven by the policies of states but by result from negotiations among the cities that are members to it” (Wehrmann *et al.* 2022, p. 115). As Dumała *et al.* (2021) emphasise, the networks themselves differ in terms of the type of cities that are members (networks of metropolises, of specialised national or regional cities), their objectives (synergy networks composed of similar cities or complementary networks with specialised cities) and scope of activities, as well as in view of their spatial range and identity.

In line with these different types of networks, the reasons for why cities collaborate in TMNs also differ. Similar to two-sided interactions, such as city-twinning, participation in TMNs is “one of the departures used by cities in aspiring for a distinct, visible, and favorable profile” (Sergunin 2014, p. 10). This may also apply to city networks, which primarily form because the member cities seek to solve problems that they cannot solve on their own and they receive access to resources, knowledge, and competencies through the networks (Dumała *et al.* 2021, Sergunin 2014). In view of city diplomacy, cities may engage in both, two-sided and multiple-sided interactions, to influence policy- and decision-making at the supra-national level (Wehrmann *et al.* 2022). As in polycentric governance, in TMNs, the member cities coordinate themselves in a non-hierarchical nature; all cities in TMNs are autonomous but implement the decisions agreed upon in the network (Dumała *et al.* 2021).

Despite all regional differences and unique challenges that cities face, not only but also in view of climate change, climate change is the most important factor shaping urban development in the Arctic due to its socio-economic and environmental impacts (Kenny 2017). Indeed, TMNs focus mostly more generally on climate change issues (Dumała *et al.* 2021, Kenny 2017). However, it is particularly the spatial identity of cities that is perceived as “the foundation of their involvement” in TMNs (Dumała *et al.* 2021, p. 13). Also in the Arctic, the activities that TMNs pursue seem to be based on the "broader Arctic identity that encompasses the entire region, acknowledging shared challenges, threats and common interests, regardless of national borders (Schweitzer *et al.* 2015)" (Kenny 2017, p. 134). As we also discussed in Chapter 2 when relating to leadership in indirect governance, the extent to which TMNs are able to achieve their goals depends "heavily on the allocation of human resources and the level of commitment of the involved cities toward becoming leaders" (Dumała *et al.* 2021, p. 13). Moreover, more formalised networks, with secretariats or other administrative bodies, "use a more wider [sic] range of tools" and "are more active" (Dumała *et al.* 2021).

The Arctic Mayors' Forum is an example of a TMN in the Arctic that was formed to connect mayors as regional leaders and local executive powers across state borders and to inform Arctic policy making. The AMF was founded "as a result of the lack of formalized involvement of local communities" (Steinveg 2021, p. 243). The AMF was officially inaugurated in October 2019, just prior to the COVID-19 pandemic, which affected its activities greatly as the mayors involved could only collaborate virtually (Wehrmann *et al.* 2022). Before the Arctic Council paused its activities in February 2022, the AMF sought to obtain observer status. Since Russia's full-scale invasion of Ukraine, the AMF paused its collaboration with mayors from Russia. A circumpolar cooperation among mayors from all Arctic countries is not to be expected as long as the war against Ukraine continues, not least because many twin cities partnerships with Russian cities in the Arctic have been terminated or suspended since 2022.

In recent years, the AMF has been increasingly formalised. The members appointed a Secretary General and the AMF has been particularly visible at Arctic Frontiers and Arctic Circle conferences (conference attendance is a so-called "soft instrument" of TMNs, cf. Dumała *et al.* 2021, p. 13). These conferences can be seen as "platforms for government officials to promote their interests in a favourable light" and "as arenas for actors to draw attention to challenges and identify these as problems that deserve attention in the problem stream" (Steinveg 2023, p. 3 and 122). Moreover, in a regional perspective, the AMF "currently stands out as the only actor representing community-driven approaches from the local level" (Wehrmann *et al.* 2022, p. 124). It remains to be seen, however, to what extent TMNs in the Arctic such as the AMF will be included in future Arctic governance, how they will be able to "foster cooperation between different levels of government [...] for addressing the specific needs of Arctic cities and their residents" (Filimonova 2023, p. 3) and whether they shape an emerging transnational Arctic society and "foster a sense of community and belonging" (Kenny 2017, p. 143).

6.3 (Transnational) Cooperation between Case Studies Cities—Insights from Interviews

As outlined in the book's introduction (Chapter 1), one of the basic premises of our project was the assumption that (transnational) cooperation in/for sustainable development in the European Arctic would already be well-developed. At the same time, we did not have idealist expectations of finding many cities here fully engaging in various partnerships to support each other in achieving sustainable urban development goals and collaborating with a focus on aligning their policies with global sustainable development arrangements and national priorities.

Our field research disclosed that this assumption needs to be adjusted and that the scale and growth of the cooperation is lower than anticipated, which contributes to a lack of coherence between the environmental, economic, and social dimensions of sustainability and limited alignment of sustainable development policies across different governance levels. In our analysis, the lack of more substantial transnational cooperation in general represents a missed opportunity to share knowledges and resources and to address shared challenges in a more unified/aligned way. Moreover, we observe that Arctic (transnational) cooperation is driven more by individual local efforts rather than by coordinated national or regional strategies.

The following section contains a qualitative analysis of the data collected from interviews in the context of two main questions, namely: (1) how does (transnational) cooperation in the European Arctic stimulate and enhance bringing together different knowledge and perspectives in the context of sustainable urban development in remote areas, and (2) does it allow to align better sustainable development policies designed at various governance levels? This analysis is supplemented by two threads that were not originally part of the research design: the impacts of the COVID-19 pandemic and the war in Ukraine.

6.3.1 How Are Multiple Knowledges and Perspectives on Sustainable Urban Development Considered in Transnational Cooperation in the European Arctic?

Following our research interest to investigate how transnational cooperation in the European Arctic promotes sustainable urban development in remote areas by bringing together diverse knowledges and perspectives, we first aimed to gauge the overall level of (transnational) cooperation in the cities we studied and how our interviewees perceive this type of cooperation. We equipped our interviewees with a glossary stating that transnational cooperation refers to the international partnership or collaboration encompassing different levels of government and administration, involving both public and private sector entities across various policy areas. None of the interviewees disagreed with this specified understanding.

Our research shows a limited scope of (transnational) cooperation in the seven European Arctic cities investigated and only moderate support for this kind of cooperation. Most of our interviewees acknowledged that the studied cities share a Nordic identity and values and often have common cultural traditions and political institutions. We argue that this recognition of a shared identity and values can serve as a foundational pillar through which transnational cooperation can thrive. It provides a strong basis for collaborative endeavours, fostering a sense of solidarity among these Arctic communities and cities. It is within this backdrop of shared Arctic identity that the potential for meaningful and impactful transnational cooperation in the region becomes most apparent.

Importantly, the interviewees often pointed out that European Arctic cities face similar challenges that require a common approach to solve them effectively. As one researcher in Tromsø (17.03.2023) explained: "Big cities and small cities across borders in the Arctic they have a lot in common, e.g., fighting attention on resources growing in southern capitals. They probably all share the same challenge of demographic challenges, shrinking populations". We perceive this common perspective and the shared imperative to address common challenges as an increasingly important argument for stronger transnational connections and intense exchanges. For example, a Greenlandic interviewee representing the NGO sector highlighted the significance of grassroots-level cooperation in the face of challenges in the circumpolar communities, which often almost naturally extends to higher levels of interaction, including foreign contacts or relations, even despite various problems associated with difficult (past) colonial experiences. Similarly, a businessperson from Akureyri emphasised the value of (transnational) cooperation, noting that pooling resources and working together advances communities by achieving common goals and enabling access to research and development that might otherwise be out of reach.

Concurrently, our research allows us to highlight the importance of a spatial dimension of (transnational) cooperation embedded in geographical imaginaries that determine the intensity and directions of this cooperation. Several interviewees underlined the remote location of their cities as a significant factor contributing to their city's peripheral status. This position was felt, for example, in Kiruna, as a tangible obstacle to continuous communication and fruitful cooperation with partners located in the southern parts of Sweden and even more so abroad. It is also worth noting that many of the interviewed citizens of Akureyri expressed their concerns or fears about the limited connectivity the town has with the Icelandic capital Reykjavik and other cities on the European continent; this was identified as a barrier for more engaged transnational cooperation.

When it comes to more constructive perceptions, we can observe that some categories of the respondents (regular citizens, city administration) prefer to cooperate primarily within the European Arctic, while others (politicians or businesspersons) prefer to view transnational cooperation as predominantly pan-Arctic. Moreover, the meaning and importance of participation in transnational cooperation differs for residents of Greenland, Iceland, Finland, and Fennoscandian states due to their unique geographical and political circumstances. Greenland's remoteness from the European Arctic, Iceland's island location, and Fennoscandian states' experiences of sectoral integration in this part of Europe can at least partially explain these differences.

6.3.2 *How Does (Transnational) Cooperation in the European Arctic Affect Policy-Making on Sustainable Development Across Governance Levels?*

Our next research objective was to investigate how (transnational) cooperation related to sustainable urban development and promotes and enables inclusive knowledge sharing in its various forms. We also aimed to examine whether sustainable urban development solutions for remote regions are developed in theses venues through co-production and partnership, incorporating input and guidance from various levels of governance. Our primary finding in this respect is that the ability of (transnational) cooperation in the European Arctic to merge diverse knowledges and viewpoints for sustainable urban development in remote regions has been only marginal. Despite facing similar urban development challenges, there is a lack of well-developed collaboration in exchanging best practices, combining different perspectives or advocating together some postulates or interests on international stages. Our research suggests that in the (European) Arctic, (transnational) cooperation with all its limitations in scope and intensity, best exemplified by the establishment of the AMF in 2019 and its subsequent actions, has been, until 2024, more strongly focused on organisational developments and promotion and dealing with external crises than on ambitious projects such as pursuing the global goals.

Our interviewees identified a few forms of institutional (transnational) cooperation, notably twin-city relationships (e.g., between Akureyri and Lahti); cross-border initiatives, such as those between Sweden and Norway, the European Union's northern Interreg programmes, and broader; and pan-Arctic efforts such as the AMF. Regarding cities' involvement in transnational exchanges, some interviewees mistakenly referred to the Arctic Council or the Arctic Economic Forum; however, they were unable to articulate the specific ways in which cities participate within these frameworks. On the one hand, this indicates or confirms the popularity of the Arctic Council, and, on the other hand, it mirrors the lack of understanding of the scope of competences of various institutions by citizens without in-depth knowledge. This may be due not only to the lack of truly multi-level coordination (orchestration), but above all to the rather marginal interest in this topic among the inhabitants of Arctic cities.

Concurrently, some interviewees, particularly those representing city administration and city councils in all cities, regularly highlighted the significance of fostering city-to-city connections, including both very well-developed bilateral contacts (twin-cities formats) and a relatively new multilateral one, the AMF (city network). These relationships were acknowledged as more than merely political, para-diplomatic or economic interactions; they were sometimes described as vital conduits for developmental and even educational exchanges. Interviewees said such partnerships can augment administrative capacities and enhance community well-being by infusing fresh perspectives and innovations. Notably, there was also a recognition of the strategic value these international alliances in a city network bring, yielding benefits and advantages not only at local but also at national and pan-Arctic levels. One interviewee from Tromsø's city administration (23.03.2023) mentioned that city networks provide local insights into broader pan-Arctic topics or national high-politics issues, which enriches the discussions and sometimes allows to keep them more practical.

Interestingly, while not yet widely recognised among our interviewees, the AMF was recalled for facilitating broader multilateral city cooperation, particularly in tackling regional pan-Arctic questions. In discussing this issue, one of the interviewees expressed the opinion that the AMF, as a multilateral structure, represents a higher and more advanced level of transnational cooperation compared to twin-city collaborations. Specifically, in our interviews, the AMF was lauded—predominantly by some city administrators involved in its works which very likely influences their views—for its focus on sustainability, ecology, green energy, and economic and social development. Moreover, it was praised for practical know-how knowledge sharing, aligning local efforts with broader global sustainable development goals. One interviewee from Tromsø's city administration (23.03.2023) mentioned that it is a platform for political innovation and bringing local urban voices into Arctic governance. Conversely, another interviewee from Akureyri's administration (14.10.2022) remarked that the AMF is not inherently political but is fundamentally about people and community engagement. Additionally, another businessperson from Akureyri (12.04.2023) criticised various forms of international collaboration as being akin to social gatherings or clubs, suggesting they do not substantially benefit communities and cities and implying that these activities are more symbolic than practical in effecting actual positive change for the communities.

Overall, our research underscores the crucial role of such connections and forms in fostering international collaboration and contributing to a sustainable future for the Arctic and beyond. In this vein, it is pertinent to note an insight from one interviewee intimately familiar with the AMF. An interviewee (city administration, Tromsø, 23.03.2023) highlighted that the participation of Russian cities as observers in the AMF ceased in 2022, a cessation attributed to the evolving international crisis and the resultant strain on Arctic cooperation. We will return to this matter further on course of this chapter (see Section 6.3.4).

Furthermore, our research highlights the human factor in transnational cooperation as it was often presented as critical. From the perspective of some interviewees (researchers), the success of (transnational) cooperation in the European Arctic is often rooted in individual relationships that emphasise the human factor beyond formal agreements. Our interviewees' insights reveal the complex nature of such collaborations, underscoring the unique challenges and opportunities they present for the future. Mutual learning, knowledge exchange, and networking are essential, and online communication platforms, including social media, that support these activities are highly valued.

In their statements on transnational cooperation in the European Arctic, only a few interviewees referred to topics or areas of such cooperation, including sustainable urban development. Examples include exchanges between Akureyri and its partner cities focusing on improving public transportation (Oulu, Finland) and addressing snow management challenges (sister city—Västerås, Sweden). It was also noted that the AMF could play a pivotal role in sustainability dialogues between cities. Interestingly, environmental and economic development issues, including infrastructure and connectivity issues, were not associated with (transnational) cooperation between Arctic cities currently or expected for the future. When interviewees mentioned these issues, they were typically in the context of other forms of international cooperation, such as foreign investment, cross-border infrastructure projects, or university collaborations. Regrettably, there was a lack of commentary on the significance of transnational cooperation in achieving sustainable development goals related to the global climate and sustainability framework. Only one individual—a local politician from Tromsø (10.03.2023) mentioned that the Paris Agreement and the 2030 Agenda are topics of discussion at AMF meetings, albeit without further elaborating.

The interviewees further indicated that the success of (transnational) cooperation among and with cities largely depends on city administration's active and committed participation. Some criticism related to this involvement was notable in several interviews, especially from individuals outside city administrations, suggesting a need for more enhanced involvement and more robust dedication of city authorities in transnational exchanges. In addition, however, some interviewees pointed out that cities not only can engage in cooperation but also receive input and react to actions from other countries and international organisations such as the European Union, as well as foreign investors and global corporations. For instance, Luleå was commended by a local politician (30.09.2022) for its regular hosting of foreign diplomats and representatives from prominent high-tech and green energy companies in recent times. Similarly, Tromsø was lauded by a researcher (17.03.2023) for its strategic position as a "gateway to the Arctic" and the presence of various entities involved in Arctic exploration and development. On the other hand, a businessperson from Akureyri (12.04.2023) expressed disappointment in the city's authorities for not taking advantage of the opportunities presented by certain international institutions located in the city, such as the secretariat of the International Arctic Science Committee. Meanwhile, a citizen from Kolari (04.03.2022) highlighted that their city lacks a clear, cohesive vision for its developmental direction and strategies.

Simultaneously, it is noted that from the perspective of city authorities, there are constraints in terms of the human resources prepared to engage professionally in developing transnational cooperation. Paradoxically, as cities intensify their involvement in international initiatives, they often face a shortage of personnel equipped or willing to sustain the increased scope and demands of such collaborative efforts.

In conclusion, it is worth mentioning the response by an Icelandic researcher who highlighted that the constructive and trustful political climate is a significant determinant of the effectiveness of (transnational) cooperation. Unfortunately, geopolitical tensions, notably the ones emerging after Russia's full-scale invasion of Ukraine in February 2022, have impeded or even have halted the opportunities for city involvement in such collaborative efforts (the withdrawal of Akureyri from the Northern Forum was cited as a case in point). However, before we present and discuss some of the consequences of the international crisis that started in 2022 for the Arctic, we provide context for how the earlier crisis—the COVID-19 pandemic—affected the European Arctic and (transnational) cooperation in the region since 2020.

6.3.3 How has COVID Impacted (Transnational) Cooperation for Sustainable Urban Development in European Arctic?

The COVID-19 pandemic has profoundly influenced both the Arctic region as a whole and in specific sub-regions such as the European Arctic. Comprehensive studies by the Arctic Council Secretariat (Arctic Council Secretariat 2020, Petrov *et al.* 2020b, Petrov *et al.* 2021b, Barik *et al.* 2022, Spence *et al.* 2022, Tiwari *et al.* 2022, Spence 2021) highlighted region-wide impact. Other researchers (Kepp *et al.* 2022, Zahl *et al.* 2023, Johansson *et al.* 2023) examined and exposed the significant importance of the pandemic for the European Arctic. This region, when compared to the North American or Russian Arctic, stands out for its relatively advanced institutional and communication infrastructures. Additionally, the European Arctic is characterised by a propensity for developing larger settlements and cities, as noted in the 2019 report by Nordregio (Nordregio 2019). These distinctive attributes played a crucial role in both the pandemic's progression and the effectiveness of the response measures implemented to control it.

The remote and often isolated communities in the Arctic, including those in the European part, encountered unique healthcare challenges (Peterson *et al.* 2023, Rapeli *et al.* 2023) during the pandemic. The pandemic severely strained the already limited healthcare infrastructure and resources, a situation detailed in studies by Nanda and Sharma (2021) and Irfan *et al.* (2022). This was particularly challenging for Indigenous communities, which often have less access to healthcare services (Dresse *et al.* 2023, Petrov *et al.* 2021a). Concurrently, some of our interviewees from Finnish cities pointed out that the pandemic's impact was somewhat mitigated by the unique urban structure of their urban spaces. Smaller urban settlements, open spaces and the larger distances between houses proved advantageous, as they naturally facilitated isolation and physical distancing within cities. Furthermore, citizens from Tromsø and Akureyri, for instance, expressed satisfaction with the pandemic-induced decrease in tourist traffic. They particularly welcomed the reduction in the arrival of cruise ships, viewing this as a positive outcome amidst the pandemic's challenges.

During the COVID-19 pandemic, inhabitants throughout the Arctic, including the European Arctic, experienced significant social and cultural impacts, as detailed in Berge *et al.*'s study (2022). Restrictions on gatherings disrupted cultural and work practices, affecting social cohesion and highlighting existing societal structural injustices, a phenomenon examined by Sigurjónsdóttir *et al.* (2021). Our interviewees also observed that the pandemic exacerbated social and familial issues, with isolation and emotional stress leading to heightened family and psychological problems. For instance, a Finnish interviewee noted that some individuals around them struggled profoundly during this period, losing hope for the future and finding even simple tasks overwhelming. More research about the mental consequences was done by Partonen *et al.* (2022). The pandemic also necessitated a shift to virtual platforms for various aspects of life, compelling institutions such as schools and universities to adapt. Some interviewees see this shift as potentially leading to enduring changes, such as the continued offering of online studies at the University of Akureyri. While this transition has benefits, including increased educational access for remote area residents and expanded opportunities for lifelong learning, it also has drawbacks, including the loss of direct student and student/professor interactions, which also affect the urban space (less students in the city will demand less services for students). Another significant area of technological growth influenced by the pandemic is telemedicine and elderly care. These remote solutions have become more prominent, addressing healthcare needs while reducing the necessity for physical travel, which is particularly beneficial in remote and isolated Arctic communities.

Our interviews conducted in Finland revealed an intriguing trend during the pandemic: a noticeable intra-country migration pattern where individuals from larger urban centres in the south of the country relocated to smaller towns in the northern parts. This movement was largely facilitated by the shift to remote work, allowing people to continue their professional jobs in southern institutions or companies while residing in the north (Ormstrup Vestergård 2022). This migration had a dual impact on the northern communities. On the one hand, these newcomers endeavoured to adapt to the local rules and customs of their new homes, showing a willingness to integrate into the existing social and cultural fabric. On the other hand, they brought fresh perspectives and ideas to these communities, particularly in areas such as sustainable development and environmental protection. As a citizen admitted (26.01.2022), most residents of the city of Kolari think mostly about their families, then about their municipality or region; the debates at the national level are distant and not often present in local discussions about sustainability. This influx of people from different backgrounds and experiences contributed then to diversifying the local discourse and potentially challenging community initiatives focused on sustainability and environmental issues.

The pandemic-induced closures of borders and travel restrictions had a significant impact on traditional activities and cultural exchanges in the Arctic (Frederick *et al.* 2021), and our interviewees shared diverse perspectives on these changes. Some expressed that the pandemic severely limited their ability to engage in usual activities and face-to-face professional interactions. They highlighted the challenges in fostering innovation and developing new ideas in a virtual environment, noting the difficulty of effecting significant change or generating creative or revolutionary solutions through online platforms like webinars. Conversely, other interviewees viewed the growth of digital communication more positively. They appreciated how it improved the connectivity of remote regions, making it easier for people in isolated areas to engage with the wider world and among each other. Additionally, they pointed to the environmental benefits of reduced air travel, which is not only costly but also has a significant ecological footprint. This shift to digital platforms, while presenting challenges in terms of personal interaction and joint development of ideas, was seen as a valuable tool in enhancing accessibility and sustainability, particularly for remote and isolated Arctic communities.

The COVID-19 pandemic underscored the critical need for enhanced healthcare infrastructure and emergency preparedness in the Arctic, a point emphasised in Saunes *et al.*'s research (2022). The pandemic also drew attention to varying national responses, such as the approach of Swedish exceptionalism, explored in studies by Christensen and Lægreid (2023) and Sandberg (2023). Furthermore, it highlighted the importance of international cooperation in addressing the challenges faced by the European Arctic, a region where issues often cross national borders, as discussed by Giacometti and Wøien Meijer (2021). While these issues were not a primary focus for many of our interviewees, a city administration representative from Akureyri did acknowledge the pandemic as a pivotal moment.

It shifted their perspective on risk assessment and threat management, revealing the limitations and vulnerabilities of existing institutional frameworks. This realisation underscored the necessity for enhanced collaboration at multiple levels of governance, extending beyond local and national boundaries to encompass a broader, transnational approach. The pandemic, therefore, served as a catalyst for re-evaluating and strengthening cooperative efforts to better address and manage the complex challenges faced in the Arctic region.

It is important to recognise that the economies of Arctic nations, particularly in European countries such as Iceland, Norway, Sweden, and Finland, are heavily reliant on key sectors such as fishing, oil and gas, and tourism. This dependency is extensively analysed in the work of Taecharungroj and Rauhut Kompaniets (2023). However, the COVID-19 pandemic brought considerable negative repercussions to these sectors. There was a notable decline in the demand for oil and gas, significant disruptions in fishing and reindeer husbandry activities (Fisktjønmo and Næss 2022) and a drastic downturn in tourism, as detailed in Paavola *et al.*s research (2023). These developments had far-reaching economic consequences, leading to adverse effects across these nations. The impact on the economy has been thoroughly examined and reported in studies by Simonen *et al.* (2022), Andersen *et al.* (2022), Wøien Meijer and Giacometti (2021), Jóhannesson *et al.* (2022) and the Nordic Council of Ministers Secretariat (2021). The pandemic thus not only affected the health and social aspects of the Arctic regions but also posed significant challenges to their economic stability and growth.

The broader impacts of the pandemic, particularly on economic issues, were a major theme in the responses we received during the fieldwork. This was especially true for interviewees in Rovaniemi, where the collapse of the tourist services market posed a significant challenge. Some interviewees highlighted the risks of relying heavily on a single industry such as tourism. They suggested that diversifying the local economy, possibly by exploring sectors such as mining, could be a more responsible approach. Interviewees from Rovaniemi, Tromsø, and Akureyri all noted the marked decline in foreign tourism as well. However, in Akureyri, this loss was somewhat mitigated by a significant uptick in domestic tourism. Another point of concern raised were the changes in the employment structure due to the pandemic. Many foreign workers, who were crucial to the tourism industry, left during this period. This situation highlighted the seasonal nature of employment in tourism and underscored the need for authorities to provide support for career transitions during such crises, allowing individuals to adapt to the changing economic landscape. One of our interviewees (business representative, Tromsø, 25.04.2023) highlighted the pandemic's significant impact on real estate, flats, and house rental service markets. In certain areas such as Rovaniemi, there was an increase in the demand for apartments. Conversely, in places such as Akureyri, many apartments, particularly those intended for short-term rentals, remained vacant. This disparity illustrates how the pandemic unevenly affected different regions and sectors. The pandemic also served as a revealing moment for existing business models, exposing their strengths and vulnerabilities in crisis situations. This led to a renewed focus on sustainable development, as it became evident which aspects truly constituted the "real capitals" of specific locations. For instance, in Luleå, there was a recognition of the need to further develop industries such as information and renewable energy technologies, sectors potentially less vulnerable to such crises and more sustainable in the long term. Furthermore, the pandemic revealed the extent of dependence that European Arctic cities have of the supply of various materials, products, services, and human resources in certain sectors, showing the importance of diversifying sources and developing more self-reliant and resilient local economies to better withstand future disruptions.

The COVID-19 pandemic resulted in a temporary decrease in human activities in the Arctic, including reductions in shipping and industrial operations, as noted by Wøien Meijer and Giacometti (2021). This decline in activity led to a short-term positive effect on the Arctic environment, manifesting in reduced pollution levels and less disturbance to wildlife, a phenomenon documented by Jóhannesson *et al.* (2022) and Baldursson *et al.* (2022). However, the long-term environmental implications are more complex and potentially concerning. The economic downturn caused by the pandemic might lead to decreased funding for environmental protection and conservation efforts. This scenario poses a significant risk to the Arctic environment, which requires sustained and adequate resources for its preservation and care. In our interviews, ecological issues in the context of the pandemic were not a primary focus, with a few exceptions. One such exception was the discussion around public transport, where the pandemic was blamed for its slower development. Another area of interest was the increased development of outdoor tourism, which could have both positive and negative implications for the Arctic environment. On the one hand, it might promote a greater appreciation and understanding of the Arctic's natural beauty and value. On the other hand, increased outdoor tourism, if not managed sustainably, could lead to environmental degradation and increased pressure on already fragile ecosystems. In view of research on the Arctic environment, the pandemic very much disrupted scientific research activities (Petrov *et al.* 2020a, Spence 2021). Many research projects, particularly those involving international collaboration, were put on hold or were cancelled, which could have long-term implications for our understanding of the Arctic environment and climate change (IASC 2020a, IASC 2020b, Uryupova 2021). None of our respondents referred to this issue directly.

The impact of the COVID-19 pandemic on sustainable urban development in the Arctic has brought two significant issues into focus. Firstly, there is the matter of social participation under restricted contact conditions. The pandemic highlighted the limitations of existing protocols for public engagement and consultation. The requirement for manual signatures on petitions in one of our Finnish case study cities, for example, emerged as a barrier, inhibiting residents' ability to participate effectively in decision-making processes. This issue underscores the need for more flexible and inclusive methods of citizen engagement, especially in times of crisis when traditional forms of participation may not be feasible. Secondly, the pandemic has brought renewed attention to the health challenges posed by the climate crisis, specifically the thawing of permafrost, and the consequent new health threats. This situation underscores the urgency of integrating sustainable development principles and goals more robustly into policy and planning.

It highlights the need for proactive measures to anticipate and mitigate the impacts of environmental changes on public health and urban infrastructure. Both issues emphasise the importance of adaptability and forward-thinking in urban planning and governance. They point to the necessity of developing strategies that not only address immediate concerns, such as those brought about by the pandemic, but also consider long-term environmental changes and their potential effects on communities. This approach is particularly crucial in the Arctic, where the effects of the climate crisis are more pronounced and can have far-reaching consequences on both the environment and human health.

In conclusion, the COVID-19 pandemic has had a comprehensive and multifaceted impact on the European Arctic, touching literally every aspect of life and society. It has significantly affected health systems, economies, environmental conditions, social structures, education, scientific research, and governance practices. The complexity and extent of these impacts reflect the interconnected nature of these various sectors and the unique challenges faced by Arctic communities. As we look forward, the long-term consequences of the pandemic in the European Arctic are continuing to evolve. While some of these effects might become more apparent over time, many of our interviewees have already identified issues that are likely to be of strategic importance in the post-pandemic era. As one citizen of Akureyri (26.04.2023) explained,

> I think we learned a lot during the pandemic and it definitely changed our way of thinking. When we make decisions, we now think more ahead and take into account all the other issues. We look at things in a broader perspective.

6.3.4 How Has the War in Ukraine Affected (Transnational) Cooperation in the European Arctic?

The war in Ukraine has had a multifaceted impact on the European Arctic, affecting various aspects of the regional multilateral and bilateral cooperation (Koivurova and Shibata 2023, Łuszczuk *et al.* 2023), regional governance (Gricius and Fitz 2022, Pic 2022), security (Wall and Wegge 2023, Strauss and Wegge 2024), economy, and society (Gole *et al.* 2022, UNHCR Nordic and Baltic Countries 29.11.2023).

The armed conflict has had a particular impact on European energy security, the green transformation, and thus on the future direction and pace of sustainable development, including in the cities of the European Arctic (Kuzemko *et al.* 2022, Osička and Černoch 2022, Zetterberg *et al.* 2022, Pereira *et al.* 2022). As it is argued in a report of the Finnish Government:

> [i]t is clear that the effects of the Russian aggression on the green and just transition in the Arctic are very complex. In the short term, the effects will be largely negative, but in the longer term, especially the energy crisis may also affect the development of more sustainable production and consumption and the acceleration of low-emission solutions both in the Arctic countries

> and globally. […] further developments of the geopolitical situation can significantly affect the attractiveness of the Arctic region as a sustainable investment destination.
>
> (cited Koivurova *et al.* 2022, p. 21)

The war that broke out in February 2022 took place during our data collection period. It came as a big shock to most of our respondents—some postponed interviews scheduled for February/March 2022; some dropped out; an official from a city administration wrote that they could not hold such interviews for security reasons and fear of Russian spies. The conflict naturally was also mirrored in the content of many of the interviews conducted in the spring and summer of 2022. In these interviews, a range of critical issues emerged, reflecting the complex and multifaceted impact of Russia's full-scale invasion of Ukraine on the European Arctic and its communities. These concerns paint a dynamic picture of a region grappling with both the immediate and long-term implications of geopolitical upheaval and seeking ways to navigate these challenges within the broader context of regional stability, economic health, and international cooperation. A key concern among interviewees was the need to support victims of the armed conflict and refugees. The welcoming of refugees to many cities in the European Arctic has significantly influenced local communities, marking a substantial shift in social dynamics. This situation is set against the backdrop of the confrontation in Ukraine and the NATO enlargement in 2023 and 2024, involving the same main actors present in the Arctic, thereby highlighting the region's growing geopolitical importance on a global scale.

Economically, according to some of the interviewees, Russia's full-scale invasion has had and is expected to continue to have a significant impact on both the European and global scale. This situation is further complicated by problems facing the functioning of the Arctic Council as the geopolitical importance of the entire Arctic region rises. The perception of the conflict's proximity varied among interviewees. For some Greenlanders, it was an event happening far from their Arctic, whereas for others in Finland and Norway, it felt much closer due to shared borders with Russia, cross-border exchanges, and a history of occupation. There was a sense of lost opportunities for cooperation with Russia, viewed by some as a significant loss. This sentiment was echoed in the suspension or severance of relations between twin-cities, such as Akureyri and Rovaniemi with Murmansk and the termination or suspension of cooperation in transnational networks such as the Northern Forum.

The war's impact on the local economy was another primary concern, with a noticeable reduction in tourism, particularly in Rovaniemi, and changes in the nationality structure of tourists, a trend observed since 2014. Interviewees also expressed worries about the future, including concerns rooted in historical experiences such as the "Winter War" in Finland during the Second World War. Challenges were noted for citizens with Russian heritage living in these cities, especially those in mixed marriages, reflecting the personal and social complexities arising from the conflict.

For some interviewees, there was an increased interest from authorities in the northern regions, potentially leading to investments in energy projects, transportation, and military infrastructure. The increasing importance of settlements in the peripheral or border regions of the European Arctic was noted, as these areas may become zones of potential expansion and exploitation by foreign countries. The need to intensify the energy transformation and develop Nordic cooperation was highlighted as a crucial response to these challenges. Uncertainties related to sanctions and their implementation were a source of concern. A strong desire for solidarity amid conflict and aggressive Russian policy was evident, with a call for more collaboration with and between Nordic states.

The war has also contributed to rising energy, heating, and other goods prices in European countries, adding to the economic strain of citizens. There was a noticeable increase in security and police control against espionage in cities. Lastly, a lack of understanding from central authorities about the problematic local situation, particularly in areas economically dependent on Russia, was a concern, as it could support pro-Russian sentiments. Overall, these insights reflect a region at a crossroads, facing unique challenges and opportunities in the wake of geopolitical shifts and seeking pathways towards sustainable development and resilience.

6.4 Conclusions

When it comes to cooperation in the European Arctic, there is a tendency towards rather passive policies and "business as usual" as the default approach, primarily due to administrative limitations, challenges in managing urban territories in harsh climates and remote places, and difficulties in attracting skilled personnel. This gap between high-level policy and local implementation impedes effective cooperation. A positive element is the openness to online communication, which was successfully implemented during the pandemic.

There is a sentiment that (transnational) cooperation is not always felt as an important measure at the local level, which is also addressed in the literature on city networks. This perception might stem from a lack of visible, immediate benefits or understanding of the value brought by such cooperation to local communities and their everyday lives. Specific initiatives such as the AMF have faced challenges due to operational issues. Before the appointment of a secretary general, the forum was described as unable to do "real work" and needing to start working more systematically. Moreover, geopolitical factors, such as the exclusion of Russian Arctic cities due to the Russian foreign agency legislation, affect the inclusiveness and representation within such forums.

There is a contrast between the local immediacy provided by networks such as the AMF and the more global perspective offered by entities such as the Arctic Council. While the former are expected to provide immediate answers and a sense of closeness, the latter offer a broader environmental and political perspective. Balancing these perspectives and integrating them into coherent policies and actions is a challenge.

As we showed in the previous chapters, in the Arctic, administrative bodies often drive cooperation that can lead to discontinuity and a lack of required constant focus on the unique challenges of Northern peripheries, especially as officials and local politicians frequently change, so do their political priorities/strategies. This highlights the need for greater political engagement and a deeper understanding of the local context, alongside more direct citizen involvement and participation in decision-making. As the experiences of the COVID-19 pandemic and the international crisis triggered by Russia's aggressive policy and war towards Ukraine have shown, external factors are also important for the scrutinised developments.

These issues encapsulate the structural, operational, and perceptual challenges facing transnational cooperation in the Arctic. As we argue in the next chapter, to overcome these obstacles, there must be a concerted/orchestrated effort to strengthen local capacity, enhance operational efficiency in cooperative forums, reconcile local and global perspectives, and encourage sustained and informed political leadership.

Acknowledgements

This chapter was written by Dorothea Wehrmann (Sections 6.1 and 6.2) and Michał Łuszczuk (Sections 6.3 and 6.4). The outline was developed by Michał Łuszczuk and Dorothea Wehrmann. The chapter was reviewed by Jacqueline Götze, Katarzyna Radzik-Maruszak, Arne Riedel, and an external expert. All authors of this book read and accepted the content of this chapter.

Notes

1 We have asked interviewees to share with us their views on (transnational) cooperation: its potential and limitations, its importance for sustainable urban development and their own experiences in this area. However, only 40 % of all interviewees responded to our questions on (transnational) cooperation. It should also be highlighted that some interviews were conducted after Russia's full-scale invasion of Ukraine on 24 February 2022, which affected how the interviewees addressed the topic of cooperation to a certain degree.

2 In the Arctic, "inhabitants live in twenty-four time zones" (Kuus 2023, p. 349), in eight states (Canada, the Kingdom of Denmark, Finland, Iceland, Norway, The Russian Federation, Sweden, the United States) with different geographic, legislative and socio-political environments; the population density in Arctic regions is comparatively low, and infrastructure and mobility are limited.

3 The Agreement on Cooperation on Aeronautical and Maritime Search and Rescue in the Arctic in 2011; the Agreement on Cooperation on Marine Oil Pollution, Preparedness and Response in the Arctic in 2013; and the Agreement on Enhancing International Arctic Scientific Cooperation in 2017.

4 As a circumpolar actor, UArctic was established amongst others to "create an Arctic learning environment" and is "strongly supported through state funding" (Nicol *et al.* 2022, p. 53).

5 As the UArctic network, IASC is also "dependent on governmental decisions and support" (Łuszczuk and Szkarłat 2022, p. 29) and plays rather an "utilitarian role" in Arctic science diplomacy (Łuszczuk and Szkarłat 2022, p. 15).

6 The Arctic Economic Council (AEC) was formed by agreement from the Arctic Council and "has produced several reports, position papers and recommendations on Arctic business development" (Menezes *et al.* 2022, p. 81).

7 Examples are the European Structural and Investment Funds programmes, Interreg programmes and the European Neighborhood Instrument cross-border programmes (for more information: Stępień and Koivurova 2017, p. 16). "In general, most EU programmes (national, Interreg cross-border and transnational) operating in the North address the specific challenges related to remoteness, sparsity and problems typical for rural areas" (Stępień and Koivurova 2017, p. 20).

8 Laruelle (2019) differentiates between the "colonial wave", which took place between the 16th century to the early 20th century, the "Soviet" wave from the 1920s to the 1980s, and the present "globalized" wave, which started in the 1960s–1970s.

9 Cities are defined in various ways. While mostly "cities are defined using criteria such as population, population density, economic function, or urban built-up area, and are usually subordinate to an administrative and jurisdictional subdivision", the definition of legal and administrative boundaries "vary significantly across countries" (DiNapoli and Juul 2020, p. 3). Also, statistical boundaries do not necessarily match with administrative boundaries and may therefore misrepresent differences. When assessing the status quo and change in cities, indicators that are based on administrative boundaries, "such as population or land area, would reflect administrative decisions, and not the actual characteristics of the urbanized area " (DiNapoli and Juul 2020, p. 3). These differences hamper the analysis and evaluation of data on cities.

10 Similar to multi-actor partnerships, the number of TMNs focusing on environmental sustainability has also grown, particularly after Agenda 21 was adopted (Dumała *et al.* 2021).

References

Andersen, T. M., *et al.*, 2022. *Nordic economic policy review 2022: COVID-19 effects on the economy in the Nordics*. Copenhagen: Nordic Council of Ministers.

Arctic Council Secretariat 2020. Covid-19 in the Arctic. Briefing Document for Senior Arctic Officials.

Baldursson, F. M., *et al.*, 2022. *COVID-19 policy responses and green recovery in the Nordic region*. Copenhagen: Nordic Council of Ministers.

Barik, J., *et al.*, 2022. COVID-19 pandemic in the Arctic and Subarctic. *In:* Shaw, R., and Pal, I., eds., *Pandemic* risk, *response, and resilience*. Amsterdam: Elsevier, 143–156.

Berge, O. K., Sigurjónsson, N., and Hansen, L. E., 2022. The consequences of the Covid-19 pandemic on Nordic cultural policy. *Nordisk kulturpolitisk tidsskrift*, 25 (2), 1–4.

Bertelsen, R. G., 2019. The Arctic as a laboratory of global governance: The case of knowledge-based cooperation and science diplomacy. *In:* Finger, M., Heininen, L., eds., *The GlobalArctic handbook*. Springer: Cham, 251–267. https://doi.org/10.1007/978-3-319-91995-9_15

Bouteligier, S., 2012. *Cities, networks, and global environmental governance: Spaces of innovation, places of leadership.* New York: Routledge. https://doi.org/10.4324/9780203106488

Christensen, T., and Lægreid, P., 2023. Crisis management of the COVID-19 pandemic: The Nordic way and Swedish exceptionalism. *In:* Cheung, A. B. L., and van Thiel, S., eds., *Crisis leadership and public governance during the COVID–19 pandemic*. Singapore: World Scientific, 289–323.

Curtis, S., ed., 2014. *The power of cities in international relations*. London: Routledge. https://doi.org/10.4324/9781315851495

Curtis, S., 2016. Cities and global governance: State failure or a new global order? *Millennium*, 44 (3), 455–477. https://doi.org/10.1177/0305829816637233

Dempsey, N., *et al.*, 2011. The social dimension of sustainable development: Defining urban social sustainability. *Sustainable Development*, 19 (5), 289–300.

DiNapoli, B., and Jull, M., 2020. Urban planning sustainability metrics for Arctic cities. *Environmental Research Letters*, 15 (12). https://doi.org/10.1088/1748–9326/abc37b.

Dresse, M. T., *et al.,* 2023. Prevalence and factors associated with healthcare avoidance during the COVID-19 pandemic among the Sámi in Sweden: The SámiHET study. *International Journal of Circumpolar Health*, 82 (1), 2213909.

Dumała, H., Łuszczuk, M., Piwowarczyk, J., and Zieliński, T., 2021. Transnational municipal networks as a mechanism for marine governance toward climate change adaptation and mitigation: Between potential and practice. *Frontiers in Marine Science*, 8, 1–16. https://doi.org/10.3389/fmars.2021.626119

Filimonova, N., 2023. *Climate change adaptation issues for Arctic and Sub-Arctic cities.* Policy Brief August 2023. BELFER CENTER for Science and International Affairs. Available from: https://www.belfercenter.org/sites/default/files/files/publication/Filimonova_Urban%20Arctic%20Adaptation%20Brief_FINAL.pdf

Fisktjønmo, G. L. H., and Næss, M. W., 2022. Consequences of COVID-19 on the reindeer husbandry in Norway: A pilot study among management staff and herders. *Human Ecology: An Interdisciplinary Journal*, 50 (3), 577–588.

Frederick, C., *et al.,* 2021. Communicating with Northerners on the absence of SARS-CoV-2 in migratory snow geese. *Écoscience,* 28 (3–4), 217–223.

Gamble, J., and Shadian, J. M., 2017. One Arctic … but uneven capacity: The Arctic Council permanent participants. *In:* Lackenbauer, P. W., Nicol, H., and Greaves, W., eds., *One Arctic*. Ottawa: Centre on Foreign Policy & Federalism, 142–156.

Giacometti, A., and Wøien Meijer, M., 2021. *Closed borders and divided communities: Status report and lessons from Covid-19 in cross-border areas.* Stockholm: Nordregio.

Gole, I., *et al.,* 2022. Economic implications of the effects of the Ukrainian War. *European Journal of Sustainable Development*, 11 (4), 17–24.

Gordon, D. J., and Johnson, C. A., 2018. City-networks, global climate governance, and the road to 1.5°C. *Current Opinion in Environmental Sustainability*, 30, 35–41. https://doi.org/10.1016/j.cosust.2018.02.011

Gricius, G., and Fitz, E. B., 2022. Can exceptionalism withstand crises? An evaluation of the Arctic council's response to climate change and Russia's war on Ukraine. *Global Studies Quarterly*, 2 (3), 1–6.

Haapala, A., 2022. Postcolonial reading of the history of urbanism in the Circumpolar North. *Barents Studies: Peoples, Economies and Politics*, 7 (1), 103–105. https://urn.fi/URN:NBN:fi–fe2022121471398

Haapala, A., 2024. Fishy windows to an Arctic city: Urban (in)visibilities of global fisheries in Tromsø. *Polar Record*, 60, e2. https://doi.org/10.1017/S0032247423000359

IASC 2020a. ASSW2021 Science Symposium Moved Online – International Arctic Science Committee [online]. Available from: https://iasc.info/news/iasc-news/591-assw2021-science-symposium-moved-online [Accessed 7 December 2023].

IASC 2020b. ASSW2020 Statement on COVID-19 – International Arctic Science Committee. Available from: https://iasc.info/news/iasc-news/546-assw2020-statement-on-covid-19 [Accessed 7 December 2023].

Irfan, F. B., *et al.,* 2022. Coronavirus pandemic in the Nordic countries: Health policy and economy trade-off. *Journal of Global Health*, 12, 5017.

Jacquier, C., 2005. On relationships between integrated policies for sustainable urban development and urban governance. *Tijdschrift voor Economische en Sociale Geografie*, 96 (4), 363–376.

Jóhannesson, G. T., *et al.,* 2022. *Arctic tourism in times of change.* Copenhagen: Nordic Council of Ministers.

Johansson, B., *et al.*, eds., 2023. *Communicating a pandemic. Crisis management and Covid-19 in the Nordic countries*. Gothenburg: Nordicom, University of Gothenburg.

Kantola, S., and Tuulentie, S., 2020. Participation in a large Arctic city – the possibilities of PPGIS for improving interaction. *Polar Geography*, 43 (4), 295–312, https://doi.org/10.1080/1088937X.2020.1767709

Käpylä, J., and Mikkola, H., 2015. On Arctic exceptionalism. Critical reflections in the light of the Arctic sunrise and the crisis in Ukraine. Available from: https://www.files.ethz.ch/isn/189844/wp85.pdf

Kenny, M. J., 2017. Urban planning in the Arctic: Historic uses & the potential for a resilient urban future. *Arctic Yearbook*. Available from: https://arcticyearbook.com/images/yearbook/2017/Scholarly_Papers/7_Urban_Planning_in_the_Arctic.pdf

Kepp, K. P., *et al.*, 2022. Estimates of excess mortality for the five Nordic countries during the COVID-19 pandemic 2020–2021. *International Journal of Epidemiology*, 51 (6), 1722–1732.

Knecht, S., 2013. Arctic regionalism in theory and practice: From cooperation to integration? *In:* Heininen, L. *et al.*, eds., *Arctic Yearbook 2013*, 164–183. https://arcticyearbook.com/images/yearbook/2013/Scholarly_Papers/8.KNECHT.pdf

Knecht, S., and Keil, K., 2013. Arctic geopolitics revisited: Spatialising governance in the circumpolar North. *Polar Journal*, 3 (1), 178–203

Koivurova, T., *et al.*, 2022. Arctic cooperation in a new situation: Analysis on the impacts of the Russian war of aggression. Valtioneuvoston kanslia. Valtioneuvoston selvityksiä No. 2022:2 https://urn.fi/URN:NBN:fi–fe2022122173093

Koivurova, T., and Shibata, A., 2023. After Russia's invasion of Ukraine in 2022: Can we still cooperate with Russia in the Arctic? *Polar Record*, 59, 1–9.

Kuus, M., 2023. The social lives of Arctic expertise, or how to do transnational networks. *In:* Heininen, L. *et al.*, eds., *Arctic Yearbook 2023*. City: Northern Research Forum, 349–360. https://arcticbook.com/arctic-yearbook/2023/2023-briefing-notes/504-the-social-lives-of-arctic-expertise-or-how-to-do-transnational-networks

Kuzemko, C., *et al.*, 2022. Russia's war on Ukraine, European energy policy responses & implications for sustainable transformations. *Energy Research & Social Science*, 93, 102842.

Lackenbauer, P. W., and Dean, R., 2020. Arctic exceptionalisms. *In:* Spohr, K. and Hamilton, D. S., eds., *The Arctic and world order*. Washington, DC: Foreign Policy Institute, 327–356. https://transatlanticrelations.org/wp-content/uploads/2020/12/The-Arctic-and-World-Order.pdf

Laruelle, M., 2019. The three waves of Arctic urbanisation. Drivers, evolutions, prospects. *Polar Record*, 55 (1), 1–12. https://doi.org/10.1017/S0032247419000081.

Laruelle, M., and Orttung, R., 2017. *Urban sustainability in the Arctic. Visions, contexts, and challenges.* Washington DC: IERES.

Łuszczuk, M., 2015. Inter-parliamentary institutions & Arctic governance. *In:* Heininen, L. *et al.*, eds., *Arctic Yearbook 2015*, 195–213 https://arcticyearbook.com/arctic-yearbook/2015/2015-scholarly-papers/131-inter-parliamentary-institutions-arctic-governance

Łuszczuk, M., and Szkarłat, M., 2022. The International Arctic Science Committee (IASC) in the changing Arctic. Between international scientific cooperation and science diplomacy. *In:* Sellheim, N., and Menezes, D. R., eds., *Non-state actors in the Arctic region.* Cham: Springer Polar Sciences, 15–41. https://doi.org/10.1007/978-3-031-12459-4_2

Łuszczuk, M., Rakowski, P., and Szkarłat, M., 2023. Poland's polar policy in the face of the Arctic Council 2022 cooperation crisis. *Polish Polar Research*, 44 (4), 339–364. https://doi.org/10.24425/ppr.2023.146739

Menezes, D. R., Couser, G., and Radkevitch, M., 2022. Highlighting businesses as key non-state actors in the Arctic: Collaboration between Arctic Economic Council and Polar Research and Policy Initiative. *In:* Sellheim, N., Menezes, D. R., eds., *Non-state actors in the Arctic region.* Cham: Springer Polar Sciences, 79–110. https://doi.org/10.1007/978-3-031-12459-4_5

Nanda, M., and Sharma, R., 2021. COVID-19: A comprehensive review of epidemiology and public health system response in Nordic region. *International Journal of Health Services: Planning, Administration, Evaluation*, 51 (3), 287–299.

Nicol, H., Beaulieu, M. S., and Hirshberg, D., 2022. The University of the Arctic (UArctic) as a non-state actor in the circumpolar world. *In:* Sellheim, N., and Menezes, D. R., eds., *Non-state actors in the Arctic region.* Cham: Springer Polar Sciences, 63–77. https://doi.org/10.1007/978-3-031-12459-4_4

Nordic Council of Ministers Secretariat 2021. *Resilience in the blue bioeconomy: What can COVID-19 teach the Arctic about the impact of crises on value chains?* Copenhagen: Nordic Council of Ministers.

Nordregio 2019. Population change in Arctic settlements [online]. Available from: https://nordregio.org/maps/population-change-in-arctic-settlements/ [Accessed 19 December 2023].

Nyseth, T., 2017. Arctic urbanization: Modernity without cities. *In:* Körber, L- A., MacKenzie, S., Westerståhl Stenport, A., eds., *Arctic environmental modernities.* City: Springer, 59–70.

Ormstrup Vestergård, L., 2022. *Strengthening Nordic cooperation on remote work and multilocality.* Stockholm: Nordregio.

Orttung, R. W., and Suter, L., 2020. Measuring urban sustainability in Arctic conditions. *In:* Orttung, R. W. ed., *Urban sustainability in the Arctic. Measuring progress in circumpolar cities.* New York, Oxford: Berghahn, 3–21.

Osička, J., and Černoch, F., 2022. European energy politics after Ukraine: The road ahead. *Energy Research & Social Science*, 91, 102757.

Paavola, J.-M., *et al.*, 2023. *Exploring domestic tourism in the Nordics: The untapped potential of the domestic tourism market in the Nordic countries and what Covid-19 pandemic taught us about how to realise it.* Copenhagen: Nordic Council of Ministers.

Palosaari, T., 2012. The amazing race: On resources, conflict, and cooperation in the Arctic. *Nordia Geographical Publications*, 40 (4), 13–30.

Partonen, T., *et al.*, 2022. Suicides from 2016 to 2020 in Finland and the effect of the COVID-19 pandemic. *The British Journal of Psychiatry: The Journal of Mental Science*, 220 (1), 38–40.

Pereira, P., *et al.*, 2022. The Russian-Ukrainian armed conflict will push back the sustainable development goals. *Geography and Sustainability*, 3 (3), 277–287.

Peterson, M., *et al.*, 2023. Public health restrictions, directives, and measures in Arctic countries in the first year of the COVID-19 pandemic. *International Journal of Circumpolar Health*, 82 (1), 2271211.

Petrov, A. N., *et al.*, 2020a. Building resilient Arctic science amid the COVID-19 pandemic. *Nature Communications*, 11 (1), 6278.

Petrov, A. N., *et al.*, 2020b. Spatiotemporal dynamics of the COVID-19 pandemic in the arctic: Early data and emerging trends. *International Journal of Circumpolar Health*, 79 (1), 1835251.

Petrov, A. N., *et al.*, 2021a. Lessons on COVID-19 from Indigenous and remote communities of the Arctic. *Nature Medicine*, 27 (9), 1491–1492.

Petrov, A. N., *et al.*, 2021b. The "second wave" of the COVID-19 pandemic in the Arctic: Regional and temporal dynamics. *International Journal of Circumpolar Health*, 80 (1), 1925446.

Pic, P., 2022. Arctic multilateralism and the consequences of the Ukrainian conflict. *Network for Strategic Analysis Policy Brief*, 20 https://ras-nsa.ca/wp-content/uploads/2022/07/Policy-Brief-20-Arctic-Multilateralism-and-the-Consequences-of-the-Ukrainian-Conflict.pdf

Pipa, A. F., and Bouchet, M., 2020. Multilateralism restored? City diplomacy in the COVID-19 era. *The Hague Journal of Diplomacy*, 15(4), 599–610 https://doi.org/10.1163/1871191X-BJA10043

Rapeli, M., *et al.,* 2023. Three Nordic countries responding to COVID-19 – Eldercare perspectives. *International Journal of Disaster Risk Reduction: IJDRR*, 84, 103442.

Sandberg, S., 2023. *The role of administrative tradition in government responses to crises: A comparative overview of five Nordic countries.* Nordicom: University of Gothenburg.

Saunes, I. S., *et al.,* 2022. Nordic responses to Covid-19: Governance and policy measures in the early phases of the pandemic. *Health Policy* (Amsterdam, Netherlands), 126 (5), 418–426.

Scott Zellen, B., 2020. Tribe-state collaboration and the future of arctic cooperation: Moderating inter-state competition through collaborative multilevel governance, from yesterday's trading posts to today's Arctic Council, 'Arctic Exceptionalism' is here to stay. *The Polar Journal*, 10 (1), 113–129, https://doi.org/10.1080/2154896X.2020.1757825

Sellheim, N, and Menezes, D. R., 2022. *Non-state actors in the Arctic region.* Cham: Springer.

Sergunin, A., 2014. *Russian sub-national actors: Paradiplomacies in the Arctic region.* Louvain-la-Neuve: European Regional Science Association (ERSA).

Sigurjónsdóttir, H. R., Sigvardsson, D., and Oliveira e Costa, S., 2021. *Who is left behind? The impact of place on the possibility to follow Covid-19 restrictions.* Copenhagen: Nordic Council of Ministers.

Simonen, J., *et al.,* 2022. The COVID-19 pandemic and regional economic resilience in Northern Finland, Norway, and Sweden – bouncing back or not? *In:* Spence, J., Exner-Pirot, H., and Petrov, A., eds., *Arctic Yearbook 2023.* Special Issue. Arctic Pandemics, 251–274.

Spence, J., 2021. Covid-19: The Arctic experience [online]. Available from: https://arctic-council.org/news/covid-19-the-arctic-experience/ [Accessed 7 December 2023].

Spence, J., Exner-Pirot, H., and Petrov, A., eds., 2022. *Arctic Yearbook. Arctic Pandemics.* Akureyri, Iceland: Arctic Portal, https://arcticyearbook.com/arctic-yearbook/2023-special-issue

Steinveg, B., 2021. *Governance by conference? Actors and agendas in Arctic politics.* Tromsø: UiT – The Arctic University of Norway. https://munin.uit.no/handle/10037/20490

Steinveg, B., 2023. *Arctic governance through conferencing. Actors, agendas and arenas.* Switzerland: Springer Nature.

Steinveg, B., Rottem, S. V. and Andreeva, S., 2024. Soft institutions in Arctic governance—Who does what?, *Polar Record*, 60, 1–7. https://doi.org/10.1017/S0032247423000335.

Stępień, A., and Koivurova, T., 2017. Arctic Europe: Bringing together the EU Arctic policy and Nordic cooperation. Finish Government's Analysis, Assessment and Research Activities15/2017 https://lauda.ulapland.fi/bitstream/handle/10024/62766/Koivurova.Timo.pdf?sequence=2&isAllowed=y

Strauss, L., and Wegge, N., eds., 2024. *Defending NATO's northern flank. Power projection and military operations*. London, New York: Routledge, Taylor & Francis Group.

Taecharungroj, V., and Rauhut Kompaniets, O., 2023. Decoding Nordic cities: Uncovering multi-level place experiences from tweets for effective city branding. *Place Branding and Public Diplomacy*, 20, 157–179 https://doi.org/10.1057/s41254-023-00313-6.

The Northern Forum 2015. *SAO Chair speaks at the Arctic Frontiers*. Yakutia, Russia: Author. Available from: https://www.northernforum.org/en/8-news/143-sao-chair-speaks-at-the-arctic-frontier

Tiwari, S., *et al.*, 2022. The second year of pandemic in the Arctic: Examining spatiotemporal dynamics of the COVID-19 "Delta wave" in Arctic regions in 2021. *International Journal of Circumpolar Health*, 81 (1), 2109562.

UNHCR Nordic and Baltic Countries 2023. If you really want to stay, you will do whatever it takes. – UNHCR Nordic and Baltic Countries. Available from: https://www.unhcr.org/neu/99687-if-you-really-want-to-stay-you-will-do-whatever-it-takes.html [Accessed on: 18.04.2024].

Uryupova, E., 2021. COVID-19: How the Virus has frozen Arctic Research [online]. Available from: https://www.thearcticinstitute.org/covid-19-virus-frozen-arctic-research/ [Accessed 7 December 2023].

Wall, C., and Wegge, N., 2023. The Russian Arctic threat: Consequences of the Ukraine war. Center for Strategic and International Studies (CSIS) Briefs. Available from: https://csis-website-prod.s3.amazonaws.com/s3fs-public/2023-01/230125_Wall_RussianArcticThreat_0.pdf?VersionId=e8h73TdoOUjdJO3Y4nOTc4v5YRmpoZad.

Wehrmann, D., 2017. Non-state actors in Arctic Council Governance. *In:* Keil, K., Knecht, S., eds., *Governing Arctic change*. London: Palgrave Macmillan, 187–206. https://doi.org/10.1057/978-1-137-50884-3_10

Wehrmann, D., 2020. The Arctic Council as a success case for transnational cooperation in times of rapid global changes? *In:* Heininen, L., Exner-Pirot, H., and Barnes, J., eds., *Arctic Yearbook 2020*. Akureyri, Iceland: Arctic Portal, 425–443. Available from: https://www.arcticyearbook.com.

Wehrmann, D., Łuszczuk, M., Radzik-Maruszak, K., Riedel, A., and Götze, J., 2022. Transnational cities alliances and their role in policy-making in sustainable urban development in the European Arctic. *In:* Sellheim, N., and Menezes, D. R., eds., *Non-state actors in the Arctic region*. Cham: Springer Polar Sciences, 113–131 https://doi.org/10.1007/978-3-031-12459-4_6

Wilson, G. N., 2019. Indigenous internationalism in the Arctic. *In:* Coates, K. S., Holroyd, C., eds., *The Palgrave handbook of Arctic policy and politics*. Cham: Palgrave Macmillan, 27–40. https://doi.org/10.1007/978-3-030-20557-7_3

Wilson Rowe, E., 2018. *Arctic governance. Power in cross-border cooperation*. Manchester: Manchester University Press.

Wøien Meijer, M., and Giacometti, A., 2021. *Nordic border communities in the time of COVID-19*. Nordregio, Stockholm: Nordic Council of Ministers.

Zahl, P.-H., *et al.*, 2023. Mortality in Norway and Sweden during the COVID-19 pandemic 2020–22: A comparative study. *Journal of Infection and Public Health*, 17 (4), 719–726.

Zetterberg, L., Johnsson, F., and Elkerbout, M., 2022. *Impacts of the Russian invasion of Ukraine on the planned green transformation in Europe*. Stocholm: IVL Swedish Environmental Research Institute.

7 Pathways towards Sustainable Development

7.1 Introduction

With this chapter, we build on the previous chapters by discussing how multi-actor cooperation and participation across governance levels may stimulate and enhance sustainable urban development in the European Arctic under and beyond the given legislation. In this chapter, we thus touch on institutional and informal practices that are informed by the theoretical premises introduced in Chapter 2 and by our data. Based on the analyses presented in the previous chapters, we indicate pathways for how governance frameworks can be advanced towards more inclusive and coherent policy approaches. With policy coherence, we refer to policy alignment in the horizontal (through local participation, see Chapter 4 on the legal foundations and Chapter 5 on the Nordic Model and participation and in the vertical sense, through cooperation, see Chapter 6). With inclusion, we address the exchanges between and within governance levels that support policy coherence.

This chapter proceeds as follows. First, we reflect on our empirical findings by bringing together the indicated challenges in policy processes[1] and ideas for a way forward derived from our "Conceptual Model: Key drivers and factors shaping local approaches to SUD" (Chapter 3). We utilise the three factors identified by the logic of our conceptual model: (1) institutional set-up and capacities, (2) actors and their relationships, and (3) priorities related to sustainable development of cities in different places. In addition, we also apply the two key drivers from the model—(a) imaginaries and (b) cooperation—to our analyses. The key drivers influence the factors: the key drivers build the basis for how visions, policies, and decisions are framed and introduced in policy-making as local approaches to sustainable urban development in the European Arctic. Second, we propose and outline three pathways towards enhanced sustainable development along different agencies. By referring to our conceptual model, these pathways present various ways that different agencies can be strengthened.

When we refer to pathways, we do not want to imply that we know what is best for the people living in the respective remote cities under investigation. Rather, we develop conceptual tools informed by theory and empirical data that can support urban development processes towards sustainable development (see also Chapters 2 and 3). These pathways suggest how the different components (factors and

DOI: 10.4324/9781003284864-7

key drivers) of our conceptual model can be changed and are open for the actors in charge to explore, adapt, and further develop. Consequently, these pathways are not final nor complete one-size-fits-all solutions. We see ourselves as external observers and hope that our outsider views complement essential insider perspectives in line with our reflections on our positionalities in Chapter 1.

7.2 Scopes for Adjustment: Identifying and Understanding the Factors and Key Drivers

With the support of our conceptual model, we analyse our data along the types of scope for adjustments—meaning the way the three factors might change to advance local approaches to sustainable urban development: institutional set-ups and capacities (factor 1), actors and their relationships (factor 2), and priorities (factor 3). Our main findings—also relating to and building on the conclusions of our previous chapters[2]—are (see Figure 7.1 for a visualisation):

1 There is an imbalance between and among actors at the local level and actors at other governance levels that influences the success of sustainable urban development and, hence, how local engagement can contribute to the global goals (imbalance of actors).
2 Local capacities to cope with urban transformation processes are insufficient (misbalance between tasks and resources).
3 Participatory processes are often overloaded in terms of content. Relevance of the decisions tied to them as well as their function for the overall municipal transformations lack thorough explanation (design of participatory processes).
4 The majority of our interviewees felt a strong domination of external actors mostly from higher governance levels or the economic sphere on urban development processes and their lives (external domination).
5 In participatory processes, economic priorities and business perspectives are often overrepresented while the perspectives of minorities are outvoted (overrepresentation of economic interests).

Building on these findings, we argue for:

a Addressing overrepresentation of economic priorities, design of participatory processes, and imbalance of actors: to ensure that no one is left behind, a more context-sensitive, inclusive facilitation of participatory processes is needed. This could help in avoiding dichotomous discussions and feelings of misunderstanding or even social exclusion. More national support for municipal decision-making is necessary to overcome many of the deficits found in the current participatory practices.
b Addressing misbalance of tasks and resources, external domination, overrepresentation of economic priorities, and imbalance of actors: local capacity building and the delegation of funds from the national to the local level could ensure more local independence and local ownership of ideas on sustainable urban development.

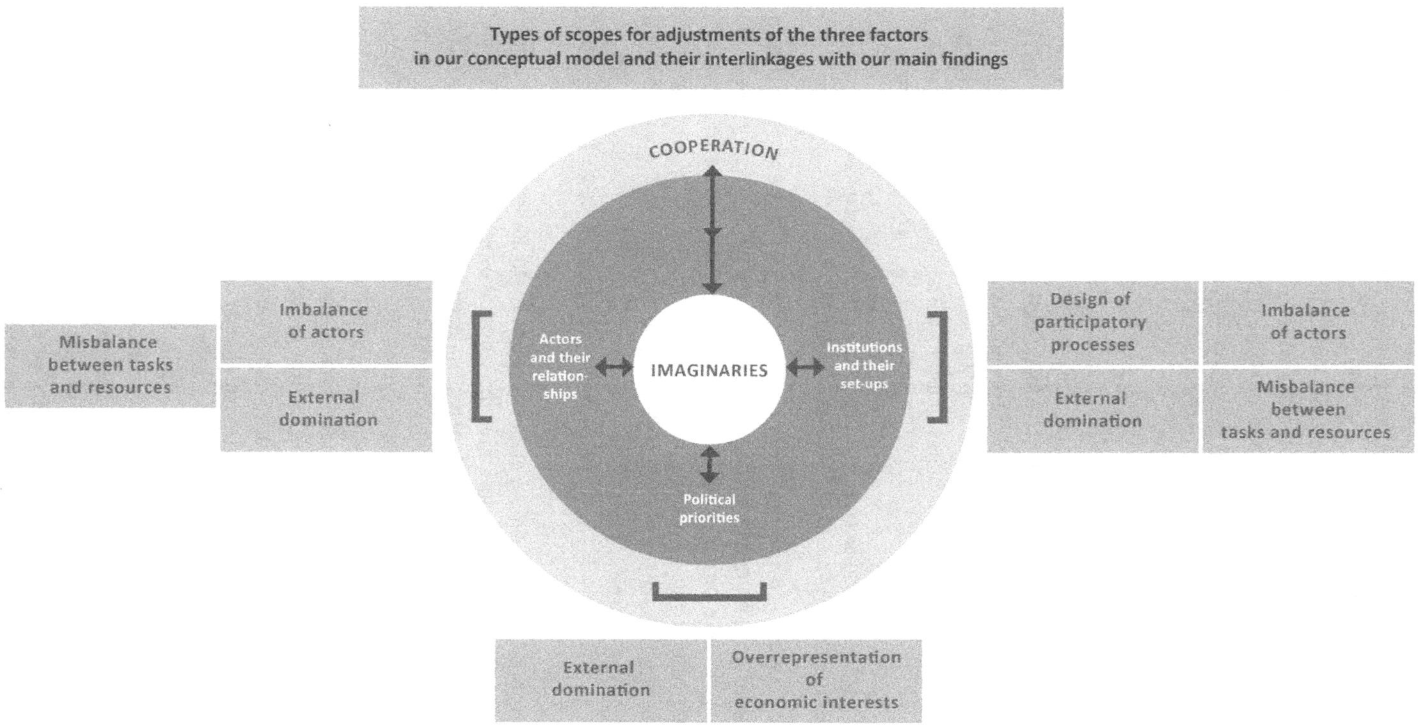

Figure 7.1 Types of scopes for adjustments of the three factors in the model and their interlinkages with our main findings. Authors work.

c Addressing design of participatory processes, imbalance of tasks and resources as well as external domination: regional Arctic cooperation and learning could be leveraged as a complementing solution to the responsibility of national governance in terms of funding and capacity building. Regional cooperation among municipal actors together with non-state actors also supports municipal capacities in the field of participatory practices while ensuring local self-determination through limiting dependencies on the national level. Regional cooperation can be further encouraged by funding schemes of actors such as the EU and the Arctic Council, and globally by the UN.

The interview data examined provides strong perceptions that sustainable urban development is hindered and indicates why urban development plans have not developed to their full potential, yet. Understanding the context for sustainable urban development to unfold in the European Arctic is a relevant first step that enables us then, in a second step, to develop ideas about how obstacles can be overcome and how potential pathways could look.

7.2.1 What Does it Take to Govern Sustainable Urban Development? Institutional Set-up and Capacities (Factor 1)

> [...] they have to ensure the services, the local services and facilities for the citizens with a very low budget because they have [a] very low population. [...] these circumstances start these vicious circles because then if people cannot have good service, they will move [to bigger cities ...] and then the population will shrink and the money for the services will decrease even more.
>
> (Researcher, Kiruna/Luleå, 20 January 2022)

For our first factor, *institutional set-up and capacities*, to understand how sustainable urban development approaches unfold in the cities investigated in the European Arctic, our data reveal that the mismatch between decision-making power and financial as well as human resources is particularly important. Our data further suggest that, in the European Arctic, the centre/periphery dynamic is quite influential for understanding the role and contribution of the local level for achieving sustainable development globally. It is influential for both, understanding the self-perception as well as the external perception of the local level in the European Arctic.

To support a better alignment of urban approaches to sustainable development vertically and horizontally through (transnational) cooperation across governance levels and participation, respectively, we deduce from our data that there needs to be more transparency for local citizens to be able to understand and act on other levels of governance. Different governance levels where decisions are taken must be better linked. We identify the local level as fulfilling a central function for the implementation of sustainable urban development. In a peripheral context, such as the European Arctic, many interviewees also stressed that citizens lack an understanding of how envisioned changes bring a benefit for them and their communities. Decisions are brought to them from external actors, while local perspectives are represented at higher governance levels only to a limited extent. This perceived lack of participation leads to frustration. To them, the European Arctic is a home, which is challenged by common representations of the European Arctic as a resource or land deposit for other regions' developments (see also Bennett 2016, Nuttall 2010).

This element of the first factor, institutional set-up and capacities, links to our findings on the mismatch between decision-making power and financial as well as human resources. Many of the analysed municipalities, for example, Kiruna, find themselves confronted with huge transformation processes and must manage vast territories, but very often lack sufficient financial and human capacity to deal strategically with these tasks in a way that contributes to sustainable development as described in global arrangements and national plans. One of our interviewees who was working in the municipal administration called their type of work "crisis management", which would allow for only little long-term, sustainable planning of urban development processes. Rather, their circumstances require decisions to be made at short notice, very often just as a reaction to the manifestation of the consequences of the climate change or of decisions made by the national government (City administration, Kiruna, 18 January 2022). Moreover, many elements of the transformation plan for Kiruna, for instance, were already decided on in 2014 when the political concept of sustainable development was not as politically prominent as today. Although long-term plans for the local level might exist that also formulate visions for the remote cities (e.g., Kiruna Kommun 2014, Kolari Kasvaa Lounnosta 2023) or the northernmost regions (e.g., in Sweden, Council for Sustainable Cities 2022), practitioners often have to make rather short-term decisions for the communities that do not necessarily contribute to long-term visions of sustainable development. Given the challenging circumstances for local administrators regarding capacity and time, it is often challenging for cities to implement sustainable development as a guiding theme for urban development. This implies a risk of leading to several non-sustainable decisions.

Some of the remote cities that we have investigated, such as Luleå, Rovaniemi, or Akureyri, fulfil a key economic role not only on a regional but very often on a national level. This economic role refers particularly to the areas of energy transition/energy security, tourism or environmental protection. However, our interviewees stressed that this economic relevance is often not reflected in the funds available for public services in the remote cities nor transferred to political relevance. In addition, tax money mostly goes south with many of the companies active in the region having offices in larger cities in the South and not in the North where the value creation takes place. The Nordic model of governance (Chapter 5) though can already provide good starting points for further developing governance approaches including local participation towards sustainable (urban) development. This is mainly due to the model's focus on citizen participation. However, our interviewees highlighted that the local government and respective legal frameworks (see Chapter 4) within which the municipalities organise their urban development projects still leave a lot of flexibility. Hence, how participatory processes are facilitated in terms of timing, participants and methods/tools lies with the municipalities. Although the frameworks legally require participation processes be organised by accompanying urban development plans, how participatory processes are implemented, for instance, who needs to be invited or who speaks as an expert about what topic is at the discretion of the responsible actors, which is often the municipal administration (see Chapter 5). Our interviewees who had already participated in different local participatory processes shared with us that particularly underrepresented, marginalised perspectives often criticise the non-inclusive setting of the processes, such as (Sámi) reindeer herders[3] in the municipalities belonging to the reindeer herding territories of the respective nation-states (such as Tromsø, Kolari, Luleå, and Kiruna). One interviewee specifically told us about participation processes that took place in different rooms for Sámi reindeer herders and the other invited actors, which made common exchanges between the groups challenging and left the participating Sámi reindeer herders with a feeling of otherness (Researcher, Kiruna, 17 January 2022).

For cases of politically more contested development projects with participants favouring conflicting land uses, in particular, for example in municipalities with greater urban transformations either ongoing or being planned (Kiruna, Kolari, or Nuuk), interviewees doubted that the current participatory framework could address the different roles of the participants involved nor influence the process and the final decision. Interviewees lamented that current rules and procedures could not protect minorities who are often outvoted by others representing bigger and more influential interest groups. A so-called balancing approach for finding compromises for conflicting interests describes another dimension of this challenge that is also explained by the consensual approach inherent to the Nordic Model of governance (Chapter 5). While from a higher-level governance perspective, a balancing approach might seem feasible for conflicting land uses,[4] from a local governance perspective, it often becomes more challenging. Different understandings of space and ideas for using land clash and create tensions at the local level (Researcher, Kiruna, 17 January 2022). Furthermore, municipalities' economic interests are tied to certain groups and sectors expressing so-called path dependencies (Researcher, Kiruna/Luleå, 20 January 2022). Thus, planning processes often either intentionally or unintentionally favour rather dominant (economic) sectors. Our data suggest that a more context-sensitive facilitation of participation processes could support avoiding such dichotomous discussions, unequal treatment of knowledge(s) and feelings of misunderstanding or even social exclusion.

While further considering the linking of vertical and horizontal alignment, the role of (transnational) cooperation became evident in our data. Several actors that are based in cities such as Luleå, Tromsø, and Rovaniemi are more involved in cooperation formats than actors based in other remote cities. The aforementioned cities have more capacity to actually work on cooperation issues that aim at bringing local perspectives together and communicate them to other governance levels in comparison to smaller municipalities (see Chapter 6). Although the decision-making power and competencies for international politics of the respective nation-states are located outside the European Arctic (with the special status of Nuuk as Greenland's capital to be considered), the Arctic regions in general stand for a long tradition of cross-border cooperation with diverse Indigenous diplomacies and municipal engagement at higher governance levels (Humrich 2017, Medby 2022, Plaut 2012, Wehrmann 2020). For many interviewees, it was less a nationally defined approach that mattered for Arctic cooperation but the people-to-people cooperation/connections following an East-West rather than North-South thinking—also challenging (post)colonial structures (Researcher, Nuuk, 16 August 2023).

Further on the issue of influence in and on participatory processes that links back to the question of capacities and resources of municipalities, our data reveal a misbalance regarding the different actors involved. Such misbalance was stressed in several interviews with municipal administrators and local governance researchers and explained by the limited amount municipal resources and great influence of certain business sections on municipal revenues (e.g., Researcher, Kiruna/Luleå, 20 February 2022). This again highlights the relevance of local capacities and resources that must be developed for local actors to be better equipped to work on challenges and to make the most out of (economic) opportunities for the communities. Based on the data, we suggest that the factor of institutional set-up and capacities could be adjusted towards more public funds being delegated from the national level to the municipalities. This would give local governance more independence for political decisions and avoid being heavily dependent on private investors for the realisation of local development projects through challenging path dependencies.

Furthermore, participatory processes often seem to be overloaded in terms of the content and relevance of the decisions tied to them. Interviewees shared that they perceive long-term processes as taking place in the background, and, at some point, they are confronted with the impacts of these processes in their everyday life. If participation processes are organised and a policy-making option is made possible, it is often not about that one decision but about the big picture and all the participation processes that did not take place (Researcher, Rovaniemi, 7 May 2021). In addition, the function of participation processes for overall municipal transformations often seem to lack thorough explanations. These observations were shared by several interviewees who took part or observed participatory processes in cities undergoing/planning bigger transformation processes, such as in Kiruna or Kolari. Many interviewees shared with us that they had had different expectations when they joined participation processes or that other participants had brought in issues that were not part of the overall design of the process but shifted its focus (Researcher, Kiruna, 17 January 2022). Participation processes can hardly serve as a panacea for solving all socio-political problems, often becoming most visible at the local level. Still, local participation should not enhance these issues either but ideally, should contribute to better local living conditions. Therefore, the objective of the participatory process needs to be clearly defined and communicated at its beginning to manage the participants' expectations. In addition, participatory processes could be more effective if the topic is rather specific and if practical and hands on suggestions/solutions can easily be devised, for, also working towards alignment, ideally, this specification should not come at the expense of losing the link to broader ongoing transformation processes and developments and across governance levels.

At this point, we identify another opportunity for designing more inclusive and transparent participatory processes that can better support and accompany urban development processes. Our interviewees stressed that local ownership of the development plans and equal access to decision-making is crucial for raising the acceptance of changes in the remote city—a recommendation also shared by urban scholars (Bilsky *et al.* 2021). Local perspectives should have the same access and speaking rights as the knowledge presented in reports by experts from outside the communities, such as scientists, businesspersons, or policy-makers from other governance levels. Several interviewees lamented too much "Southern" influence, meaning influences coming from outside the European Arctic from capitals and bigger cities in the South, on urban development processes in their communities (Researcher, Kiruna/Luleå, 20 December 2021 and Researcher, Kiruna/Luleå, 20 January 2022). This element of local ownership and ultimately local self-determination is also highly relevant against the backdrop of postcolonial debates on (economic) developments in the European Arctic (Junka-Aikio 2022, Schweitzer 2022). We see a strong link to capacity building in the European Arctic for people to have access to higher education in the region and to have the possibility to stay in their home communities (Huynh and Lidmo 2022, von Redecker 2023). Human capacity building is impacted greatly by local/regional educational possibilities and job prospects (Hirshberg and Petrov 2014). The regional universities in the remote cities under investigation[5] fulfil a key role in that regard, as well as examples of university cooperation.[6] These institutions and networks provide education from the region for the region, allowing people to get a university degree without the need to go South and on topics relevant for the region. Furthermore, cross-regional cooperation between the universities facilitated by university networks can provide increased capacity for Arctic research at large through also addressing social, cultural and technological needs in the North.

The described Southern influence on European Arctic life and the relevance of capacity building from within the European Arctic further illustrate the centre/periphery dynamic between municipalities in the remote region and urban centres/capitals in the South, often referred to by our interviewees. However, this dynamic not only exists between the "North" and the "South", but interviewees also indicated dynamics between peripheries with more powerful urban centres and the more peripheral areas in the remote North. With view to our case study cities, we can see these dynamics particularly around the regional hubs of Luleå, Rovaniemi, Tromsø, and Akureyri. This dynamic takes place in a multi-level context and is strongly linked to the imaginaries (key driver 1) that shape the different governance levels as well as influence how these levels interact with each other.

In sum, the factor of institutional set-up and capacities could be adjusted in the sense that municipalities must be equipped with the adequate resources (financial, human, time, local self-determination) to design development projects in a more long-term and sustainable way (Lidmo *et al.* 2020). Another element of this factor implies more guidance by the national level for local governments because these local developments often contribute to national objectives and fulfil nationally defined needs or priorities (in the case of reindeer herding and mining, for instance). This guidance is not to be misunderstood with external domination but rather it should equip remote cities with the relevant capacities and tools to implement often-nationally-defined goals. A third, complementing element that could adjust this factor is more exchange across governance levels to discuss the interplay between them and to better ensure local perspectives being represented at higher governance levels on issues affecting the local level. To create more opportunities for local perspectives to be represented at higher governance levels would further counteract feelings of external domination. Local actors would have the opportunity to bring in their knowledges, perspectives, and experiences and, hence, could already flag certain local limitations and needs for capacity building during exchanges with actors from higher governance levels.

7.2.2 Actors and their Relationships for Sustainable Urban Development: Finding a Balance? (Factor 2)

> It has been crisis management for 3–4 years now. […] On Friday when I go home, I leave the work at work. […] but I don't know what's happening on Monday.
>
> (Administration, Kiruna, 18 January 2022)

The misbalance between local and national influence on local development is also visible in view of the dimension of the actor and their relationships. In all cities under analysis, our interviewees stressed that the municipalities' often rather limited local capacities cannot cope with implementing the political decisions made at higher governance levels with influence beyond the local level. This situation creates challenging working conditions for municipal administrators as well as local politicians who must find local solutions for problems that were very often brought to them from outside the region but need to be managed locally (Administration, Kiruna, 18 January 2022 and City Council, Luleå, 8 May 2023). In the European Arctic, for instance, development projects in the area of renewable energy, battery storage, or mining that all contribute to nationally defined goals are examples. They are perceived as external and economic benefits that often do not stay in the communities.

To better understand local perspectives in their diversities,[7] interviewees pointed to the differences between newcomers and residents who lived longer in the area; between Indigenous peoples pursuing traditional livelihoods, other residents who are working jobs in sectors in conflict with such traditional livelihoods; and between generations. In the case of Kolari, for example, during the COVID-19 pandemic, many people moved from southern, urban areas to the more remote Finnish Arctic to enjoy nature and tranquillity.[8] In our interviews, these so-called "newcomers" identified themselves as residents of Kolari. Other interviewees were local people living in Kolari for generations. Some want to push forward the reopening of the mine and do not see these newer residents as legitimate people from Kolari who should have a say in decisions on Kolari's future (see also Chapter 5). The discussions about the reopening of the mine not only follow along the divide of new and long-term residents but also run along generations and within families. People who used to work in the former mine have quite nostalgic memories about the "old times" and see the mine as *the* opportunity for Kolari to regain economic splendour (Researchers, Kolari, 28 January 2022). In contrast, people working in the tourism sector, including both longer-term citizens and those who have recently moved there, are worried that the reopening of the mine would endanger their business model of selling holidays in a healthy environment.

Further, the influence of different actors on urban development processes and their accompanying participatory processes already elaborated under the previous factor indicate a relevant actor dimension, too, that also needs to be addressed under the factor of actors and their relationships. As shared by our interviewees (Researcher, Kiruna/Luleå, 16 December 2021 and Researcher, Kiruna/Luleå, 20 December 2021) and analysed by other scholars (e.g., Garbis *et al.* 2023) the focus of urban development projects in processes steered towards sustainable development often is strongly directed at economic questions. This also influences participatory processes with an often-perceived bigger influence of economic actors in comparison to lesser influential actors representing other social/interest groups. This economic focus is additionally strengthened by greatly varying conditions and opportunities for the different actors to participate in terms of financial, time and professional resources. For instance, business representatives can invest time and personnel to prepare their participation. In contrast, representatives of other groups need to participate in their free time without a professional background. For the latter, participation is an extra effort while for the first it is part of their job and paid time. Another element that can again enhance the influence of this actor group appears in single-industry municipalities, such as in the case of Kiruna. Several interviewees criticised the strong influence on urban development by the state-owned company operating the mine. In this case, the company is not only quite influential but also even pays for the urban transformation as well as the accompanying participation process. In addition, its position as a state-owned company indicates connections to the private and governmental sphere. This describes a quite powerful actor in comparison to the other involved actors in the participation processes.

Following an actor-centred, relational approach in light of this factor, even though there are many actors at the local level contributing to policy-making on pursuing sustainable urban development, our interviews focused mainly on city administrations. Despite the focus on municipally led initiatives towards vertical alignment through transnational cooperation and horizontal alignment through participation processes at the local level, informal processes at both the transnational and local levels in the European Arctic also play a role in the development of local approaches to sustainable urban development. However, some of our interviewees highlighted the important role of non-governmental engagement, for example, through grass roots initiatives and volunteer work particularly in the field of social cohesion (e.g., Citizen, Luleå, 5 April 2022, Politician, Tromsø, 20 March 2023, Researcher/citizen, Nuuk, 29 April 2022) and on higher governance levels, the idea for a pan-Arctic self-organisation of civil society actors.[9]

Building on these elements of the factor on actors and their relationships, we suggest adjusting this factor by introducing measures for counterbalancing the influence of the economic sector, which we see again linked to the analysis for the previous factor in terms of adequate financial resources for local governments. These financial resources for the local level could also support in counterbalancing the often, rather unequal, distribution of power between the governance levels. Secondly, another dimension for adjusting entails a power-sensitive facilitation of the participatory processes accompanying the urban development projects to avoid confrontations between different social groups. Thirdly, the remote city and local politicians should ideally support civil society engagement at different governance levels financially and institutionally to allow for these perspectives to be included in decision-making as well (see Chapter 5). Fourthly, to overcome the problem that politicians and administrators often seem to not work hand-in-hand, we propose paying more attention to how the complex relationships between both can be improved, which, in line with research on public management, seems to be inseparable in the context of urban development.

7.2.3 What are the Local Priorities for Sustainable Urban Development? Who Defines Them and How Are They Negotiated? (Factor 3)

> [...] in how far is this development paradox that we often discuss when it comes to the Arctic is this something that is discussed differently on the local level in view of the economic sectors, and perhaps also regionally and in how far are there references, for instance, to the climate agreement and to the whole discussion about climate change.
>
> (Researcher, Tromsø, 13 January 2022)

By looking at Factor 3, priorities, we see again many functional links to the previous elaborations on the first and second factors informed by our data. Linked to the local ownership of development plans and ideas (*institutional set-up and capacities*, Factor 1) as well to the diversities of local actors (*actors and their relationships*, Factor 2) our data deduced from the interviews suggest that these conditions ask for ongoing local discussions on what sustainable development means and entails for the communities. As the residents and generational perspectives change so does the living environment; moreover, the local understandings of sustainable development are not static, which influences long-term plans and visions for the community. Interviewees shared that new people bring in novel ideas and different experiences that can also change local discourses. This challenges the view of a homogeneous local community as well as any clear lines between long-term and recently moved citizens and their right to have a say in decisions on local development. Moreover, interviewees also highlighted that long-term residents might change their perspectives in terms of what creates good living conditions for people while contributing to sustainable development. Hence, we see a potential here for the local level to develop formats for facilitating co-creative ongoing discussions on sustainable development as well mechanisms to feed these local perspectives in a meaningful way into urban development projects (Lindberg *et al.* 2020).

A local framing of good local living conditions supports these suggestions further. Our interviewees shared that local developments are not only about contributing to a bigger picture of global, regional, and national initiatives, but should also create tangible benefits for the communities. This links back to the analysed local/global connection under the first factor on institutional set-up and capacities and also to the centre/periphery dynamic analysed under the same factor. Economic developments must create value for places to avoid any feelings of exploitation.

In this regard, the need for local representation at other governance levels becomes evident, for example, through cooperation. This relates to the balancing approach and the need to find solutions for conflicting topics that are discussed at the local level but whose scope is beyond it. Such cooperation must create value for the local level and ideally avoid closed cooperation settings. In our data, we see that the engagement of municipalities in transnational city alliances in which they connect with one another often falls in the context of non-state actors as these activities go beyond their core tasks (Wehrmann *et al.* 2022). This different role gives remote cities more flexibility, but actors might use this flexibility to the citizens' disadvantage.

In the case of the (European) Arctic, one interesting case is the Arctic Mayors' Forum (AMF) that has been already introduced in the previous chapter (Chapter 6). When going back to our theoretical elaborations on the need for leadership in forms of indirect governance such as orchestration (see Chapter 2), the example of the AMF can be illustrative. Structural leadership can be given by the respective nation-states as well as the EU.[10] Cities as members of this multi-actor partnership provide entrepreneurial and cognitive leadership by bringing in experiences and knowledge from the local level and contexts. The role of an exemplary leader could be fulfilled by the AMF. As such, the AMF could advance the forum's inclusivity and support its orchestrating abilities—convening, agenda-setting, providing assistance and synchronising activities (cf. Abbott *et al.* 2012). Hence, it could become more institutionalised. For example, the forum could apply for funding at different levels and actors, such as the EU with its cross-border programmes and use these funds to support smaller municipalities to join the forum. Accordingly, this factor could be adjusted by developing measures that enable smaller municipalities to join transnational cooperation formats to have their perspectives represented as well by contributing to a better understanding of diverse local priorities. Further engagement in and institutional development of formats like the AMF could contribute to both—a better vertical and horizontal alignment of policies that support sustainable (urban) development. For the cities and their citizens, this entails a stronger voice and representation through the AMF at the transnational level following the "We are the Arctic" narrative. Among the members of the AMF as a multi-actor partnership, the AMF setting allows for not only sharing and exchanging knowledges and experiences but also for co-creating solutions and policy-responses.

Given that priorities can change, processes must be flexible to address such varying priorities and constantly allow for (re)defining local priorities. This does not contradict a long-term plan for a community as long-term plans can also allow for changes. For a vertical alignment, formats such as the AMF and others could fulfil a central role in terms of orchestration of local priorities that are expressed in different policies, initiatives and perspectives (Abbott *et al.* 2012, Abbott *et al.* 2016).

7.2.4 How Do Key Drivers Come in?

Transitioning to the key drivers—Imaginaries and Cooperation—these follow a different logic than the factors of the conceptual model that we have analysed before. The key drivers cannot be changed directly. Rather, they must be indirectly influenced by adjusting the factors. While the adjusted factors can develop or hamper cooperation, imaginaries rather describe diverse internalised mind-sets and mental maps. The latter do not have to be adjusted/changed, but need to be reconstructed as such imaginaries that might guide and shape policy-making in and for the region; they must also be critically (re)considered and reflected as such.

There are, on the one hand, historically traditionalist external imaginaries about the European Arctic. They are mainly defined by what the European Arctic is not or how it is less and/or different to other places outside the European Arctic. On the other hand, we find newly (re)emerging, self-developed and self-told imaginaries. Both types of imaginaries are shaping the region's self-understanding as a whole and its specific localisations, its developments as well as questions of vertical and horizontal alignment for sustainable development. The different imaginaries can conflict with each other, but they can stimulate each other and can display direct links and interlinkages or exist entirely separated from each other. As indicated above, and with a view to the key drivers, we can identify multiple links between the different dimensions that inform each other. This helps us to better understand the context for sustainable urban development to unfold, which is needed to develop pathways in the first place.

From 'We Need the South' to 'The South Needs us' (Imaginaries as key driver 1)

> [T]he people who live in [the] Arctic [...] in these [...] peripher[ies], these people want to have better lives, they want to have better houses, they want to have better chances. But then of course you change and the people who come in from the cities, they tend to think 'oh no, you should stay original. You should stay in contact with nature'. Tourists expect us to live in line with this romantic idea, which never was true. Of course, we are changing constantly, but there is a kind of pressure on that, we should stay the same.
>
> (Researcher, Akureyri, 27 May 2022)

This quote is illustrative of many of the elements describing the imaginaries, the first key driver in our model because it touches upon the difference/tension between self-imaginaries and external ones as well as their intersections. Further, it reveals a specific type of relationship between the (European) Arctic and the "rest", the more Southern regions. Here, this is particularly insightful against the backdrop of the Arctic's remoteness and colonial past (and postcolonial present) (cf. Shadian 2018, p. 276).

The imaginaries can be linked to several dimensions of the characterisations we have elaborated on in the factors on institutional set-ups and capacities, actors and their relationships, as well as priorities. We identify strong links particularly with the centre/periphery dynamic described under all factors of our model. For instance, against the background of this centre/periphery dynamic between Southern regions and the European Arctic, as well as the local ownership of ideas, many interviewees related to the need to develop definitions of a good life in the European Arctic without comparing it to life in the South. Understandings of development and its objectives might be different between people(s), communities and places, which stand out above all in the European Arctic due to its remoteness, different livelihoods and colonial past/postcolonial present (Lawrence 2014, Normann 2021).

Likewise, associated with the centre/periphery dynamic, rather traditionalist imaginaries of the environment describe the European Arctic as having vast lands and being rich in natural resources. To avoid the feeling of being highly determined by political decisions taken elsewhere favouring the European Arctic as a land and resource deposit to be developed for actors from other regions, self-determined communities must be equipped with strong local government capacities. Further, they must have access to decision-making processes at higher governance levels. Accordingly, remote cities would be enabled to decide on what kind of development is locally wanted and beneficial in social, environmental and (long-term) economic terms (Researcher, Kiruna/Luleå, 20 January 2022).

For a remote city to benefit from the economic development pushed by external actors and to distribute the revenues locally, many interviewees stressed the need for these companies to have an office in the remote city and pay local taxes. If headquarters are located outside the remote city, often somewhere in the South, the tax revenues are not paid in the remote city where the value creation took place. In cases of land-intense and sometimes environmentally harming economies, such as the promotion of renewable energy or mining, local tax revenues must also cover the environmental (and social) costs for the remote city to avoid externalisation of costs and exploitation. Another dimension of this issue is fly-in-fly-out workers and employees who work in the municipalities in limited-time economic sectors, such as mining. Without a permanent residence, these employees do not pay income tax in the remote cities where the value creation took place but in their home cities elsewhere (Researcher, Kiruna/Luleå, 20 December 2021). Further, this type of work has social implications since families are either living somewhere else or only temporarily housed in the remote city. This creates challenging circumstances for the remote cities when it comes to service provision and planning.

These elements also speak to what is often referred to in the (European) Arctic as a development paradox, "a phenomenon capturing the intertwined spheres of economic development and environmental protection in Arctic development" (Łuszczuk *et al.* 2022, p. 30). Specifically, "the ADP [Arctic Development Paradox] illustrates the normative trap of prioritising access to resources and socio-economic development at the expense of the environment, or vice versa, protecting the environment by limiting economic prosperity" (Łuszczuk *et al.* 2022). This development paradox also informs a political paradox that influences participatory processes accompanying urban development plans (Politician, Rovaniemi, 20 February 2022 and Researcher, Tromsø, 13 March 2023). Dichotomous discussions about "protecting the environment" versus "promoting economic development" or "respecting the rights of Indigenous peoples" and how these three intersect/hinder each other transfer to participation as these dichotomies build the underlining dynamics of several urban development processes. In other words, it is unlikely that one participatory process solves a paradoxical, political challenge as described by the ADP that is a multi-level and multi-issue phenomenon. However, the ADP impacts many developmental decisions taken at different governance levels, which demands constant and critical reflections by the actors involved in development processes.

Another element that can change the imaginaries of the environment and shows strong links to sustainable urban development is the urbanisation of the European Arctic as a result of urban development processes and colonisation. The phenomenon of urbanisation can act as a political tool changing the dynamics between different municipalities and within municipalities in the (European) Arctic as a remote region (cf. Researcher, Kiruna/Luleå, 20 January 2022). The urban development of certain places can raise the remoteness of others and can canalise decision-making power with new peripheries being constructed. Urbanisation is further intertwined with colonialism—particularly in the Northern part of Norway, Sweden, Finland and in Greenland. Urban settlements were built on traditional Indigenous land and still challenge (urban) imaginaries of the Arctic (Laruelle 2019, Nyseth 2017).

When further referring to imaginaries in the European Arctic, another element shaping both self-understandings and the characteristics attributed from outside the region is the belief or image of a pure and natural environment. Interviewees shared that this image remains unspoilt despite urban development processes and behaviours, which are not necessarily sustainable. Against this backdrop, interviewees also stressed the role of tourists who come to the European Arctic for its nature and beautiful environment, which again strengthens the perception of the locals of the space (Researcher, Akureyri, 27 May 2022). The need to care about "sustainable" urban development does not seem a priority given the already perceived clean, green and natural environment in the European Arctic.

In contrast to this element on hindering external and self-attributed imaginaries, we see another relevant element in our data that is informed by the shifts induced by climate and environmental change. Our case study cities see themselves already confronted with these today and call for local solutions (Administration, Kiruna, 18 January 2022). This links to new European Arctic identities being constructed with a potentially new dynamic between the North and the South. Given the political urgency to rebuild and transform energy systems to become more sustainable and independent, the EU as well as the respective nation-states in the European Arctic have defined objectives. These aims concentrate on the promotion of renewable energy and domestic mining for critical minerals to increase electric mobility and data storage. In all three areas—renewable energy, mining for critical minerals and data storage—the European Arctic fulfils a central role for Europe to achieve its goals. Our data suggest that the municipalities in the European Arctic have noticed this new interest in their region and that they see a potential to change certain dynamics between the North and South in that regard. Such a new dynamic would exert influence on sustainable urban development as well (City Council, Luleå, 8 May 2023 and City Council, Luleå, 3 March 2022). While the traditionalised imaginary would rather define a dependence of the Northern parts on the Southern parts, new European Arctic identities would describe it the other way around: "the South needs us".

Cities in the European Arctic: Towards More (Transnational) Cooperation? (key driver 2)

> We're trying to engage ourselves much more in our Arctic future in our framework conditions, we're understanding much more that if we want to have a say and if we want to have ownership and responsibility for our own lives, our own futures then we do need to understand that our existence is not just local, regional, or national. It is international and it is pan-Arctic.
>
> (Administration, Tromsø, 24 March 2023)

> There isn't any other organisations like the AMF, it is not political, it is about the people.
>
> (Administration, Akureyri, 14 October 2022)

Our data showed fewer activities from the remote cities under investigation in the area of (transnational) cooperation than we expected. As we argue in Chapter 5, cooperation does not work as an independent factor but is rather shaped by the three factors explained before. By understanding the limited activities as missed opportunities to share experiences and knowledges, we want to look for potential orchestrated efforts to strengthen local capacities. These efforts could enhance the overall operational efficiency in cooperation formats, align local and global perspectives and, ultimately, encourage more sustained and informed political leadership. In doing so, we want to contribute to a more nuanced understanding of cooperation between the remote cities under investigation.

As already derived in the previous chapters in this book, local perspectives represented at higher governance levels contribute to a stronger vertical and horizontal alignment towards enhanced sustainable development. For that, the municipal level must have resources to become involved and to have actual access to the transnational level. Our interviewees highlighted the role of this engagement in identity-building for the municipal actors—operating internally and externally. However, linked to the construction of new peripheries and the exclusiveness of cooperation formats, such as the AMF, interviewees also critically asked about the local value of these cooperation formats. Since comparatively bigger municipalities have more capacity to get involved, as an engagement that goes beyond their core tasks, cooperation formats can become exclusive settings. Other voices stressed the potential of cooperation while elaborating the need either for new formats that could also promote local perspectives at higher governance levels despite the municipal level (e.g., through a more institutionalised organisation of Arctic civil society actors at different governance levels) or for reforming existing ones (such as the AMF). The role of actors at other governance levels for supporting the establishment of new formats and reforming existing ones was also strengthened and relates to the wording, for instance, of the current EU Arctic Policy from 2021 that prominently mentions the AMF (European Commission 2021, p. 13). Only quite recently has the EU established the "Arctic Urban-Regional Cooperation" (AURC) programme by gathering 15 municipalities from all over the Arctic (except for Russia) from Canada, Finland, Greenland, Iceland, Norway, Sweden, and the United States (European Union External Action (EEAS) 2024). The AMF was involved in establishing the new programme and was able to include all of its member municipalities into the AURC. Framed as a complementary programme to the AMF, the European Commission, the European External Action Service, and the Directorate-General for Maritime Affairs and Fisheries fund this initiative.

Mainly due to external influences—overall, the COVID-19 pandemic and Russia's full-scale invasion of Ukraine—multi-actor partnerships such as the AMF were bound by crisis management with little time and capacity left to support internal projects among its members (Chapter 5). The pandemic hit the AMF only shortly after its establishment, and Russia's full-scale invasion of Ukraine had a far-reaching impact on the cooperation in the region overall, shaping the relations between the members of the AMF with Russian cities that were excluded. To be more adaptable and feature a forward-thinking approach, the AMF, as well as other transnational formats in the Arctic and beyond, learned from these external shocks.

After having elaborated how our data can speak to the components of the conceptual model introduced in Chapter 2 by identifying and understanding factors and key drivers, the next section presents the pathways we have developed that build on these analyses.

7.3 Pathways towards Enhanced Sustainable Urban Development

Building on the elaborations and characterisations for the factors and key drivers, in this section we suggest three potential pathways towards enhanced sustainable urban development, linking vertical and horizontal alignment (see Figure 7.2). While the previous section took a local governance perspective, we also want to address the structures and relationships between governance levels with the pathways. Hence, our pathways identify the links between local and other governance levels by reflecting on the current division of tasks. Accordingly, the pathways follow a multi-level governance approach, here particularly a polycentric governance that understands specialisation, division of tasks, subsidiarity as well as addressing local-regional circumstances and community preferences as tools to create more efficient policies (Morrison *et al.* 2017 cited Morrison *et al.* 2023, see also Chapter 2). For that, access to finances, knowledge(s), and social and basic services is needed, which demand coordination, integration, and participation. The analysis of our data along the conceptual model developed in Chapter 3 identified capacity building and leadership as key for the development of policies towards enhanced sustainable (urban) development. With our pathways, we address the questions of linkages/ leadership (structural, entrepreneurial, cognitive, and exemplary; Chapter 2.1) and orchestration, namely of who fulfils the role of a convener, an agenda-setter, a provider of assistance, and a synchroniser of activities (Abbott *et al.* 2016).

We differentiate the pathways along the dimension of agency to build on one of our research insights on the imbalance of power of the different actors involved in policy-making on sustainable (urban) development. We frame the pathways (see Figure 7.2) as all ideally leading to the same aim, which we define as advanced participation and alignment through local participation (Pathway 1), transnational formats (Pathway 2), and the design of global agreements in the context of sustainable (urban) development and regional cooperation (Pathway 3). Accordingly, we develop local pathways to sustainable development that could also be scaled-up to feed into the broader established scenario research that aims to develop and model pathways combining qualitative narratives and quantified trajectories to a sustainable future (cf. Soergel *et al.* 2021). The pathways can intersect and cumulate. We write them in present tense to show their character as alternate realities. In Pathway 1, we describe responsibilities that lie with the local and national level and share recommendations for a reform of the Nordic Model of governance to a Nordic Model 2.0. For Pathway 2, we identify the local and transnational level as key and focus here on cooperation and the building/strengthening of multi-actor platforms (MAPs), particularly, city networks. Pathway 3 brings the national governments together with the regional and global level into the centre while addressing the nexus of the role of global agreements and regional cooperation. It relates back to the concept of orchestration and the need for concerted efforts for the global agendas to be successful. National governments act as intermediaries who bring

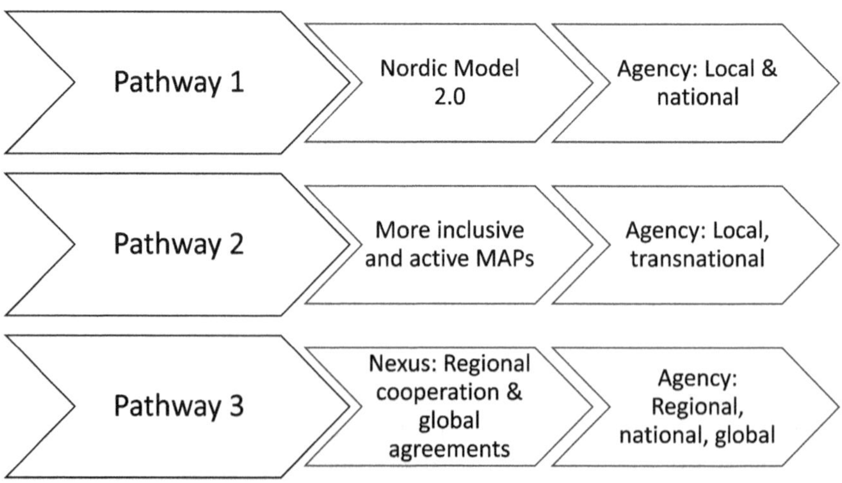

Figure 7.2 Pathways towards sustainable urban development. Authors work.

(ideally) local perspectives into decision-making processes at the global level. At the same time, MAPs, such as the AMF, fulfil exemplary leadership and act as orchestrators. The municipalities provide entrepreneurial and cognitive leadership. Further, the level of the EU, relevant for the European Arctic, can provide structural leadership through the (financial) support and facilitation of (regional) cooperation. Since all these levels of governance interplay, some elements of the pathways also intersect and overlap. Hence, the pathways are not to be understood separately, but different remote cities might choose or combine different pathways or prioritise one pathway over the other(s) as they are also targeting different governance levels.

Pathway 1 *Nordic Model 2.0 for more local agency*

The Nordic Model is reformed and further developed to the Nordic Model 2.0, which involves devolution and localisation. As the former Nordic Model with its division of tasks led to challenges with implementation at the local level, following the reform, local governments and actors are included earlier in the decision-making at higher governance levels. This way, local capacities can be addressed from the beginning. The model is adapted to the needs of current societal challenges, gives the local level more agency, and, hence, equips (remote) cities with better capacities to fulfil a variety of tasks and competencies. For this, more public funds are directed from the national towards the local level to allow remote cities to have the financial means to cope with urban transformations. At the same time, the national governments support the local level with more guidance for sustainable urban practices towards enhanced local capacities to cope with sustainable urban development, particularly through a focus on more innovative and inclusive participatory processes. The reform especially addresses remote cities to avoid the construction of peripheries.

Pathway 2 *Towards more inclusive and active MAPs at the transnational level*

MAPs, such as the AMF, become more inclusive for remote cities to join through the support of bottom-up initiatives of member cities and also with cities outside the platforms. The platforms provide framework-setting activities and (re)distribute funds to counterbalance different capabilities and capacities of the cities involved (including remote cities). Beyond local engagement at the transnational level, non-state actors' engagement is becoming more aligned (e.g., by following Humrich's idea on an Arctic Civil Society Council presented at the Arctic Circle Assembly 2022 in Reykjavík) with bringing together local perspectives that are not orchestrated by a (trans)municipal actor, as well as with building a different kind of legitimacy that complements (or even supports) local governments' initiatives and vice versa.

Pathway 3 *Strengthen the nexus of regional cooperation and the design of global agreements*

This pathway describes developments towards a more regional approach to facilitating policy alignment across regions and governance levels. For the achievement of the global goals, the international community (meaning the United Nations, UN), in particular, and national governments hold agency. As an orchestrating entity, it is in the hands of the UN and its member states to ensure that the agreed-upon global goals are pursued. The national governments (as intermediaries) who signed the agreements and instruments have a special (two-way) responsibility in this regard because they must report to the UN (more explicitly to the High-Level Political Forum, HLPF), and they need to support the local governments (as targets) in their pursuit of sustainable urban development. In this way, it is the national governments' responsibility to coordinate and facilitate the knowledge exchanges and capacity building needed to achieve the global goals. Regional cooperation must complement and further support national and global engagement by bringing together local perspectives. This way, competencies in urban development stay with the local level.

After having developed these three pathways towards sustainable urban development, in the next chapter, we open our scope of research to other remote regions while building on our findings on sustainable urban development in the European Arctic as a case study. We refer back to our conceptual model and reflect on its applicability to other remote regions.

Acknowledgements

This chapter was written by Jacqueline Götze. The outline was developed by Jacqueline Götze, Michał Łuszczuk and Dorothea Wehrmann. The chapter was reviewed by Michał Łuszczuk, Katarzyna Radzik-Maruszak, Arne Riedel, Dorothea Wehrmann and two external experts. All authors of this book have read and accept the content of this chapter.

Notes

1 By following the exploratory questions, which guided our research, we coded our qualitative data along the questions of: (i) how sustainable urban development in the Arctic can be steered more effectively in alignment with local and global policies (cf. Chapter 4), (ii) how multi-actor and participatory approaches may stimulate or enhance sustainable urban development in the European Arctic under and beyond the given legislation (by particularly referring to our elaborations in Chapter 5 and Chapter 6), (iii) how sustainable development could be practised ideally and how local and transnational cooperation can be enhanced to align sustainable development, and (iv) how and under what conditions transnational cooperation stimulate and enhance bringing together different knowledges and perspectives in the context of sustainable urban development in remote areas.

2 To recall our findings of the previous chapters along the various factors and key drivers, in Chapter 4 we identified the role of different governance levels for implementing the global agreements; in Chapter 5, we saw gaps between formal and informal role of institutions as well as diverse policy priorities; and in Chapter 6, we highlighted the relevance of but limitations for cross-border cooperation in the European Arctic.

3 In Norway and Sweden, reindeer herding is only practised by Sámi people. In Finland, reindeer herding can be practised by ethnic Finns and Sámi people (Sarkki *et al.* 2021, p. 274). Please also consider our reflection on the diversities of different actor groups as well as Indigenous peoples in the European Arctic.

4 In the remote cities under investigation, these conflicting land uses are, for instance, mining (in case of Kolari and Kiruna), (mass) tourism (Rovaniemi, Akureyri, Tromsø) and (renewable) energy promotion facilities (Luleå) on the one hand, and nature-based tourism, reindeer herding and other traditional land uses on the other hand.

5 These are the Luleå University of Technology (LTU), the University of Akureyri, the University of Lapland in Rovaniemi, the Arctic University of Norway (UiT) with one of its campuses in Tromsø and the University of Greenland.
6 These are, for instance, the University of the Arctic (UArctic) or the Arctic Five (network between LTU, University of Umeå (Sweden), UiT, University of Lapland and University of Oulu (Finland)).
7 Please consider Chapter 1 on our reflections about "who are the locals".
8 In addition, almost 50% of the people living in the Nordic states have access to a second home in the countryside, informing urban-rural linkages with people from urban areas impacting smaller municipalities (Slätmo and Kristensen 2021).
9 Humrich's idea of an Arctic Civil Society Council presented at the Arctic Circle Assembly 2022.
10 See also the EU's recently established programme for Arctic cities, "Arctic Urban-Regional Cooperation (AURC)" programme (European Union External Action (EEAS) 2024).

References

Abbott, K. W., Genschel, P., Snidal, D., and Zangl, B., 2012. Orchestration: Global governance through intermediaries. Available from: https://papers.ssrn.com/sol3/papers.cfm?abstract_id=2125452 [Accessed 21.08.2024].

Abbott, K., Genschel, P., Snidal, D., and Zangl, B., 2016. Two logics of indirect governance: Delegation and orchestration. *British Journal of Political Science*, 46 (4), 719–729. https://doi.org/10.1017/S0007123414000593

Bennett, M. M., 2016. Discursive, material, vertical, and extensive dimensions of post-Cold War Arctic resource extraction. *Polar Geography*, 39 (4), 258–273. https://doi.org/10.1080/1088937X.2016.1234517

Bilsky, E., Moreno, A. C., and Fernández Tortosa, A., 2021. Local governments and SDG localisation: Reshaping multilevel governance from the bottom up. *Journal of Human Development and Capabilities*, 22 (4), 713–724. https://doi.org/10.1080/19452829.2021.1986690

Council for Sustainable Cities 2022. Visions in the North: The future in municipalities in northern Sweden (Swedish original: Visioner i norr: Framtiden i kommuner i norra Sverige). Available from: https://arkdes.se/en/designed-living-environment/practice-based-research/visions-in-the-north/ [Accessed 21.08.2024.]

European Commission 2021. *Joint communication to the European Parliament, the Council, the European Economic and Social Committee and the Committee of the Regions. A stronger EU engagement for a peaceful, sustainable and prosperous Arctic*. Brussels: European Commission. Available from: https://eeas.europa.eu/sites/default/files/2_en_act_part1_v7.pdf [Accessed 21.08.2024.]

European Union External Action (EEAS) 2024. 15 towns and cities starting new cooperation across the Arctic [Press release]. Available from: https://www.eeas.europa.eu/eeas/15-towns-and-cities-starting-new-cooperation-across-arctic_en [Accessed 21.08.2024.]

Garbis, Z., *et al.*, 2023. Governing the green economy in the Arctic. *Climatic Change*, 176 (4), 33. https://doi.org/10.1007/s10584-023-03506-3

Hirshberg, D., and Petrov, A., 2014. Education and Human Capital. *In:* Larsen, J. N., and Fondahl, G., eds., *Arctic human development report. Regional processes and global linkages*, Copenhagen: Nordic Council of Ministers, 348–399.

Humrich, C., 2017. The Arctic Council at twenty. Cooperation between governments in the global Arctic. *In:* Conde, E., and Iglesias Sánchez, S., eds., *Global challenges in the*

Arctic region. Sovereignty, environment and geopolitical balance. London: Routledge, 149–169.

Huynh, D., and Lidmo, J., 2022. *Nordic overview of national support initiatives in urban planning*. Copenhagen: Nordic Council of Ministers.

Junka-Aikio, L., 2022. Whose settler colonial state? Arctic Railway, state transformation and settler self-indigenization in Northern Finland. *Postcolonial Studies*, 26 (2), 1–23. https://doi.org/10.1080/13688790.2022.2096716

Kiruna Kommun 2014. In-depth master plan for Kiruna central town 2014 (in Swedish: Fördjupad översiktsplan Kiruna centralort 2014). Available from: https://kiruna.se/bygga-bo--miljo/kommunens-planarbete/oversiktsplaner/fordjupad-oversiktsplan-kiruna-centralort-2014.html [Accessed 21.08.2024.]

Kolari Kasvaa Lounnosta 2023. Municipal strategy of the municipality of Kolari 2023–2027 (In Finnish: Kolarin kunnan kuntastrategia 2023–2027). Available from: https://kolari.fi/media/kunta-ja-paatoksenteko/tiedostot/kuntastrategia-2023-2027.pdf [Accessed 21.08.2024.]

Laruelle, M., 2019. Postcolonial polar cities? New Indigenous and cosmopolitan urbanness in the Arctic. *Acta Borealia: A Nordic Journal of Circumpolar Societies*, 36 (2), 149–165. https://doi.org/10.1080/08003831.2019.1681657

Lawrence, R., 2014. Internal colonisation and Indigenous resource sovereignty: Wind power developments on traditional Saami lands. *Environment and Planning D: Society and Space*, 32 (6), 1036–1053. https://doi.org/10.1068/d9012

Lidmo, J., Bogason, A., and Turunen, E., 2020. *The legal framework and national policies for urban greenery and green values in urban areas. A study of legislation and policy documents in the five Nordic countries and two European outlooks*. Nordregio Report 2020. https://doi.org/10.6027/R2020:3.1403-2503

Lindberg, M., *et al.*, 2020. Co-creative place innovation in an Arctic town. *Journal of Place Management and Development*, 13 (4), 447–463.

Łuszczuk, M., *et al.*, 2022. Governability of regional challenges: The Arctic development paradox. *Politics and Governance*, 10 (3), 29–40. https://doi.org/10.17645/pag.v10i3.5341

Medby, I. A., 2022. A "peopled" account of political agency in the Arctic: Professional practice and people-to-people participation. *The Geographical Journal*, 189 (3), 412–421. https://doi.org/10.1111/geoj.12486

Morrison, T. H., *et al.*, 2017. Mitigation and adaptation in polycentric systems: Sources of power in the pursuit of collective goods. *WIREs Climate Change*, 8 (September/October), 1–16. Available from: https://onlinelibrary.wiley.com/doi/full/10.1002/wcc.479 [Accessed 12 September 2023].

Morrison, T. H., *et al.*, 2023. Building blocks of polycentric governance. *Policy Studies Journal*, 51, 475–499. https://doi.org/10.1111/psj.12492

Normann, S., 2021. Green colonialism in the Nordic context: Exploring Southern Saami representations of wind energy development. *Journal of Community Psychology*, 49 (1), 77–94. https://doi.org/10.1002/jcop.22422

Nuttall, M., 2010. Oil and gas development in the North: Resource frontier or extractive periphery? *The Yearbook of Polar Law Online*, 2 (1), 225–243. https://doi.org/10.1163/22116427-91000036

Nyseth, T., 2017. Arctic urbanization: Modernity without cities. *In:* Körber, L.- A., MacKenzie, S., and Westerstahl Stenport, A., eds., *Arctic environmental modernities: From the Age of polar exploration into the era of the Anthropocene*. City: Palgrave Macmillan, 59–70.

Plaut, S., 2012. "Cooperation is the story"—Best practices of transnational indigenous activism in the North. *The International Journal of Human Rights*, 16 (1), 193–215. https://doi.org/10.1080/13642987.2011.630498

Sarkki, S., Heikkinen, H. I., and Löf, A., 2021. Reindeer herders as stakeholders or rights-holders? Introducing a social equity-based conceptualization relevant for Indigenous and local communities. *In:* Nord, D. C., ed., *Nordic perspectives on the responsible development of the Arctic: Pathways to action.* Cham (Switzerland): Springer Nature, 271–292.

Schweitzer, P., 2022. Between resource frontier and self-determination: Colonial and postcolonial developments in the Arctic. *In:* Finger, M., and Rekvig, G., eds., *Global Arctic: An introduction to the multifaceted dynamics of the Arctic.* Cham: Springer International Publishing, 51–367.

Shadian, J. M., 2018. Navigating political borders old and new: The territoriality of Indigenous Inuit governance. *Journal of Borderlands Studies*, 33 (2), 273–288. https://doi.org/10.1080/08865655.2017.1300781

Slätmo, E. and Kristensen, I., 2021. Urban-rural linkages: An inquiry into second-home tourism in the Nordics. *In:* Bański, J., ed., *The routledge handbook of small towns*. London: Routledge, 218–231. https://doi.org/10.4324/9781003094203

Soergel, B., *et al.*, 2021. A sustainable development pathway for climate action within the UN 2030 agenda. *Nature Climate Change*, 11 (8), 656–664. https://doi.org/10.1038/s41558-021-01098-3

von Redecker, E., 2023. *Bleibefreiheit (freedom from lead)* (3. ed.). Frankfurt a. M. Germany: S. Fischer Verlag.

Wehrmann, D., 2020. The Arctic Council as a success case for transnational cooperation in times of rapid global changes? *In:* Heininen, L., Exner-Pirot, H., and Barnes, J., eds., *Arctic Yearbook 2020.* Akureyri, Iceland: Arctic Portal, 425–442.

Wehrmann, D., *et al.*, 2022. Transnational cities alliances and their role in policy making in sustainable urban development in the European Arctic. *In:* Sellheim, N., and Menezes, D., eds., *Non-state actors in the Arctic*. Cham, Switzerland: Springer, 113–131.

8 Sustainable Urban Development in Remote Regions beyond the Arctic?

8.1 Introduction

The aim of this book is to better understand why policies for sustainable urban development (SUD) are not aligned across governance levels in remote regions.[1] While the previous chapters focused specifically on the European Arctic, this chapter broadens the geographical perspective to remote regions in general. Under consideration of the often-stressed unique environmental, political, and socio-economic context in the Arctic (Arctic exceptionalism),[2] the purpose of this chapter is threefold[1]:

First, we compare how the challenges and opportunities for advancing SUD in the European Arctic identified in the previous chapters differ from other remote regions—specifically in view of local participation and transnational cooperation. We compare evidence from studies that focused on other remote regions with the findings from our case studies in the European Arctic (Akureyri, Kiruna, Kolari, Luleå, Nuuk, Rovaniemi, and Tromsø). Here, we focus specifically on the question of how, in the context of SUD, local participation and transnational cooperation are interpreted by scholars in other remote places. To recall, the 2030 Agenda, the Paris Agreement, and the New Urban Agenda prioritise governance across multiple levels (1) and inclusive, participatory multi-actor approaches (2) as means to pursue the global goals. The purpose of this section is, therefore, to explore the extent to which local approaches to SUD in other remote regions elsewhere are shaped by policies decided at other levels of governance and the extent to which local approaches to SUD are based on inclusive, participatory structures.

Second, based on the findings of this comparison, we discuss the implications for the scientific debate on SUD in remote regions in general: we compare the findings from the literature on SUD in other remote regions with insights shared by our interviewees (based on our qualitative content analysis) and research results presented by scholars at a hybrid workshop, which took place in May 2022 and served as a sounding board of our preliminary research results (cf. Summary Report, SUDEA 2022 and Chapter 1). In this section, we also consider the main findings on participation and transnational cooperation introduced in the previous chapters.

Third, we sketch out the limitations in terms of transferability of our model on how to adjust the components of SUD (as introduced in Chapter 3) in particular: we identify the differences in the academic debate on remote regions, highlighting

DOI: 10.4324/9781003284864-8

gaps for further research and potentials and limitations regarding the transferability of our model beyond the (European) Arctic. In particular, we discuss how SUD is pursued in the light of the place-sensitive factors introduced in Chapter 3:

1 institutional set-up and capacities,
2 actors and their relationships, and
3 political priorities.

As explained in Chapter 3, we argue that these factors affect (1) how local actors engage in policy-making processes for advancing SUD, (2) the policy-making on sustainable development in general, and (3) the understanding and approaches to sustainable development in remote regions. By considering the assumption that remoteness remains the basis of imaginaries that firmly define, frame and fill SUD (key driver 1, Chapter 3) and that cooperation stimulates the pursuit of global goals (key driver 2, Chapter 3), we discuss to what extent these three factors can be subject to change for advancing SUD in remote regions in general (i.e., in terms of the role attributed to cities, the links between cities and other actors involved in policy- and decision-making processes, and the way in which leadership is exercised).

This chapter is based on an extensive literature review, a qualitative content analysis of information shared in interviews on our case cities between December 2020 and June 2023[3] and research results presented during a hybrid expert workshop on "Advancing Sustainable Urban Development in Remote Regions". We considered literature from other remote regions irrespective of the countries they are located in and are aware of the need to be context-sensitive to understand the evolution of local approaches to SUD. However, the purpose of this chapter is not to trace the evolution of local approaches to SUD in other remote regions. Instead, we compare whether or not similar or other factors are considered crucial for how SUD is approached in other remote regions. We are also aware that the research presented in this chapter is limited by the keywords that we used in our literature review (see below).

8.2 Local Participation and Transnational Cooperation in Other Remote Regions

In the SDG literature, research on cities is highly relevant, but cities in remote regions are still rarely considered (Hawken *et al.* 2021, Vale *et al.* 2023). The majority of articles that we identified by using the specific set of key words are published in journals with an explicit urban focus (e.g., Sustainable Cities and Society, Urban Studies, Urban Research and Practice) and relate to cities located in Europe, Asia, and North America. Those studies that focus on cities that are characterised by some kind of remoteness refer mostly to "peripheral" or "remote" areas, regions, locations, cities, or megalopolises (Scholvin 2021, Vale *et al.* 2023, Xiao *et al.* 2023).

In this section, we considered studies published in the databases SCOPUS, OPAC, and DEBIS[4] to explore the practices and relevance ascribed to local participation and transnational cooperation in other remote regions. We draw from studies that relate to remote places with urban characteristics in the broadest sense and refer to the terminology used by the authors. To examine if and how knowledge exchanges across governance levels and regions take place, we relate first to studies that explore local approaches to SUD and if/how they are shaped by policies at other levels of governance. Second, we evaluate the relevance ascribed to local participation and transnational cooperation and scrutinise to what extent local approaches to SUD are based on (formalised) inclusive, participatory structures.

8.2.1 Local Approaches to Sustainable Development

There is a broad consensus to recognise disparities between remote and non-remote cities, including in terms of sustainable development (Coenen *et al.* 2021, Frank and Hibbard 2017). These disparities are also reflected in the terminology used in in-depth studies that categorise remote cities as distant "resource peripheries", "gateways", or as "sites of innovation and creativity" (Leane *et al.* 2021, Patel 2022, Scholvin 2021). The latter two illustrate that the functions ascribed to cities are often similar, irrespective of their remoteness. However, in the literature on remote and/or peripheral regions that focuses explicitly on sustainability transitions, "peripherality is mostly associated with power imbalances, (lack of) political representation and voice", which "shape and hinder the adaptive potential of peripheral regions" (Vale *et al.* 2023, p. 8 and 10). In view of energy transitions, this power imbalance is, for example, perpetuated in resource peripheries, which are places "where value is extracted and captured by actors located elsewhere" (Coenen *et al.* 2021, p. 224), leaving environmental and social costs to the local communities (externalisation of costs). Power imbalances are also visible in view of the "multi-scalar embeddedness of transition policy", which is, according to Vale *et al.* (2023), one key dimension addressed in the literature on peripheral regions. Yet, only few studies investigate systematically how local approaches to SUD are shaped by policies at other levels of governance in remote regions.[5] While in the field of urban planning, for example, it is acknowledged that the concept of a compact city has been applied in policies that aimed at improving "urban transport sustainability" around the world for decades (Smith and Barros 2021), systematic evidence on how such concepts are transferred across governance levels in remote places is still missing. Such evidence is, however, needed to concretise how a "holistic policy-mix" can be deployed, without neglecting place-sensitivities. In this way, Vale *et al.* also demand more discussion on "concrete types of policies and interventions [that] are to be favoured to help the navigation of peripheral regions in the turbulent waters of sustainability transitions" (Vale *et al.* 2023, p. 11).

In the context of climate change mitigation, in contrast, studies that focus on smaller cities in rural areas in countries such as Germany and the United States have stressed that rural areas and smaller cities and towns have lower capacities, less population and fewer young and educated people, and that local governments are often poorer. As a consequence, they "apply less for climate funding [and] also receive lower amounts", seem less engaged in climate mitigation activities, and are considered “unlikely pioneers” for climate change mitigation (Ken 2019, and Homsy 2018 cited in Zeigermann, Kammerer, and Böcher 2023, p. 898 and 910, Marschall *et al.* 2021). Such studies thus imply that local climate mitigation approaches are shaped rarely by policies and initiatives at the national or regional levels. This corresponds to the observation that “political jurisdictions and policy actions tend to intervene in a limited and bounded space” (Vale *et al.* 2023, p. 11). Interestingly, however, these studies indicate that the composition of local parliaments, and particularly the number of left and green seats, did not explain “variation in the local climate action as it does at the national level”. Instead, they hint at the influential role of “local policy entrepreneurs” and “local civic capacity” (Bulkeley and Kern 2006, Zahran *et al.* 2008 cited Zeigermann *et al.* 2023, p. 908–909).

Few studies, however, explore systematically how citizens encourage policy innovation (Harvey-Scholes *et al.* 2023), and, to our knowledge, none of them focuses on remote regions (an exception being Lindberg *et al.* 2020). Irrespective of the remoteness of cities, this observation supports the demand for more formalised structures to advance knowledge consideration and transfer (Vale *et al.* 2023), such as so-called “shared” or “third spaces” (Patel 2022). Such spaces shall facilitate the co-production of knowledges “in the face of mounting urban complexity and uncertainty” as has often been emphasised in research on transition pathways for sustainable development and “function to reconfigure power relations embedded in traditional systems of knowledge production” (Patel 2022, p. 376). We will relate to such formalised structures further next.

8.2.2 Formalisation of Inclusive, Participatory Structures

In the literature on sustainability transitions, formalised structures facilitating knowledge transfer such as “third spaces” are conceptualised differently and encompass different entities. In view of cities, for example, the smart city vision and respective city branding (H.-J. Wang 2023) builds on the ideal to incorporate the perspectives of different actors in city development processes, similar to the understanding of multi-actor partnerships and multi-stakeholder platforms (Chapter 2). Cities themselves, however, can also be seen as polycentric spaces as urban spatial structures “comprise[...] a set of organizational rules that connect behaviours and interactions between urban morphological and functional subsystems” (Yu *et al.* 2022). In this way, for example, the concept of “portals for knowledge exchange” places the dynamic exchange between individuals and institutions at its centre and highlights “the need to include processes for anchoring institutional learning”, and the “crossing [of] institutional boundaries to co-produce knowledge” for sustainability pathways and transformations (Patel 2022, p. 385). Such portals can be established not only at the local level but also at other governance levels.

Interactions among citizens, collaborations between citizens and elected representatives, and also interactions between different cities are often described as affecting progress towards achieving the SDGs (Harvey-Scholes *et al.* 2023,

Xiao *et al.* 2023). The evidence presented on formalised participatory structures, however, is surprisingly little in the literature on remote regions. One reason might be that respective structures still need to be formalised. Another might be that interactions require connectedness, and connectivity is often described as a problem in rural and remote regions, as for example, studies on broadband connectivity in Brazil and the development of airborne transport in China show (Cavalcante *et al.* 2021, Hao *et al.* 2020). Connectedness and network approaches, however, are seen as "a fundamental attribute to challenge peripherality" (Vale *et al.* 2023, p. 10) but different to city networks; only few studies have focused on networks connecting remote or rural areas (Y. Wang *et al.* 2023). As research on Chinese cities shows, the construction of transportation, communication, and information networks in particular are seen as improving network connectivity between cities and advancing "inter-city synergistic effects" for their sustainable development (Xiao *et al.* 2023). However, such connections also seem to spur "a certain clubbing effect", with growing cities being more strongly connected and shrinking cities being less connected (Fu *et al.* 2022, Groth *et al.* 2023, Zhou *et al.* 2023, p. 1261). This assessment is also supported by evidence from other regions, where resource wealth or the concentration of economic or political activities turn remote places into "gateways" (Groth *et al.* 2023, see also Chapter 6 on the construction of new peripheries).

In line with centre-periphery relationships, these, however, often "appear to prosper at the expense of peripheral locations, whose role in the corresponding networks is reduced to basic functions" (Scholvin 2021) and are thus also weakened in view of inter-city cooperation, infrastructure (Kosai and Yamasue 2021), and mobility services (Groth *et al.* 2023). Other studies also point to the challenges that particularly "non-metropolitan" regions face in "balancing inward and outward relationships", which are amplified by sociocultural and religious drivers (Raufirad *et al.* 2017), by strong regional identities and a greater focus on intra-local networks. Less "locked-in actors", particularly those that are connected "to the outside world" obtain powerful roles and are seen as "important agents behind sustainability transitions" (Vale *et al.* 2023, p. 10).

While participation, connectivity, and coordinated (local, national, and transnational) network-approaches are generally seen as beneficial for sustainable development, we found surprisingly little evidence on formalised structures in studies on remote regions. Despite significant differences in sub-national realities, however, such formalised structures that facilitate local participation and cooperation across governance levels and regions are widely seen in the literature on remote regions as key for more holistic policy-making on SUD. In this way, the co-production of knowledge is essential (Patel 2022). When considering the concept of orchestration (see Chapter 2), foremost orchestrators but also intermediaries would be in charge of establishing such structures to coordinate knowledge exchanges and advance participation. As we explain in Chapter 2, different techniques can be used to establish and bring such structures to life, which also strengthen linkages (connectivity) among all actors involved (cf. p. 19ff.).

8.3 To What Extent Are Processes Related to Sustainable Urban Development in the European Arctic Different from Other Remote Regions?

"[P]eriphery is not a fixed, objectively geographical or economic condition: it is shaped by social, political and economic relations between different regions and actors, and depends on the specific functions at stake" (Vale *et al.* 2023, p. 3). This quote emphasises two important aspects that we repeatedly found in research on cities and on remote regions: the relational embeddedness of places (1) and the dynamic nature of cities and their development that is shaped foremost by interactions among people and less by geography (2).

In what follows, we discuss to what extent the findings from the literature on SUD in other remote regions relates to insights from the European Arctic. Without neglecting the disparities among cities, which result from the two aspects highlighted above, we focus specifically on the second aspect, the dynamic interactions among humans and, here, on participation and transnational cooperation for SUD. We concentrate on power imbalances and the lack of formalised structures, which were repeatedly highlighted in the literature cited above and by our interviewees and scholars. We do not provide a systematic comparison of the different approaches to SUD in the European Arctic and in other remote regions, as this is beyond the scope of this chapter and—as described above—the data available from other remote regions proved to be very limited.

8.3.1 Power Imbalances across Governance Levels, Among Cities and Individuals

Power imbalances affect the development of remote cities in a number of ways. Despite differences between cities,[6] the literature cited above highlighted in particular the lack of political representation of remote cities and the lack/limited inclusion of diverse perspectives in policy-making at the local level and across governance levels as a result of power imbalances. The limited consideration of local experiences and perspectives on sustainable development across governance-levels was also stressed in our fieldwork (Chapters 5 and 7) and during our workshop with experts who related to the Arctic and to other regions. Scholars highlighted, for example, that the concept of SUD might be too normative and that it seems to be forced on a social reality while the plurality and different social realities are not reflected in SUD processes (SUDEA 2022, with an overview on this debate).

Similar to the findings on other regions, power imbalances between cities reinforce the peripherality of less connected cities in particular. Connectivity relates here not only to infrastructure but also to the inclusion in networks and the building of local coalitions ("clubbing"). As has been found in other remote regions, cities with growing populations and economies (such as Tromsø and Rovaniemi) are more strongly connected than those affected by outward migration (such as Kolari). In the European Arctic and also in other remote regions, there is evidence that limited connectivity also reproduces centre-periphery relations between remote cities with negative effects on political representation and economic development. Both economic and political inequalities are further often related and can reinforce each other, as for example research on the representation of Sámi people illustrates (SUDEA 2022, Götze 2024). Also, our interviewees often positioned the (European) Arctic as a region grappling with economic stagnation. This stagnation is attributed to several factors: the peripheral locations of Arctic communities, the harshness of the Arctic climate, and the implications of climate change, as well as demographic challenges such as sparse population density. These elements underscore the region's struggle to foster (sustainable) economic growth. In addition, the Arctic is often portrayed as a target of external exploitation and resource extraction. This viewpoint highlights the dynamics of power and vulnerability, where local Arctic communities find themselves in unequal confrontations with multinational corporations or governmental entities. These external actors are often depicted as imposing exploitative practices on the region, seeking concessions or changes that prioritise (short term) economic gain outside the region over local welfare and environmental and social sustainability. During the interview, an Icelandic researcher encapsulated this sentiment by noting, "poor places are more vulnerable for offering the space, resources for sell" (Researcher, Akureyri, 27 May 2022).[7] Moreover, the interviewees from Luleå also pointed to a burgeoning trend in Northern Sweden, where cities and towns vie for the attention of investors. This competition has been exemplified by the establishment of a green energy megasystem in Norrbotten, which stimulated such local rivalry. However, in the eyes of some interviewees, this green transformation is not without its complexities. It represents not only a record-breaking level of investment but also the re-industrialisation of the region and a strain on a relatively small labour market. Moreover, this shift has contributed to a gradual decline in other economic sectors.

Such developments have led to a significant alteration in the socio-spatial structure of Norrbotten, a region traditionally viewed as peripheral within Sweden. The emergence of dominant development centres within the region has been accompanied by the decline of intra-regional peripheral areas, reshaping the economic and social landscape. A researcher with ties to both Luleå and Kiruna has observed that these transformations redefine the "northern identity" of the Swedish North (Researcher, Luleå, 16 December 2021). As a result, conflicts over land use and ownership intensify and highlight the complex interplay between economic development, environmental sustainability, and sociocultural dynamics in the region. These changes underscore the multifaceted challenges and opportunities that come with the pursuit of green industrialisation in traditionally remote and peripheral areas. At the same time, many citizens in the European Arctic feel neglected by national policy-makers, particularly due to the peripheral status of Arctic communities as articulated by a resident of Akureyri.

> [It] is good to bear in mind in the context is—maybe one thing and this is not only Icelandic or specific to Akureyri, this also pertains to the Arctic—there is this tension, the centre periphery tension which animates a lot of the politics, a lot of the discourses that go around. So basically, people in the

> periphery like Akureyri they feel themselves always at the receiving end of the globalized forces, whichever way conceived, it passes through Reykjavik.
>
> (Researcher, Akureyri, 14 September 2022)

Such neglect is believed to pervade various policy domains, reinforcing centre-periphery tensions and contributing to a broader sense of alienation across the Arctic. This sentiment suggests that the issues facing Arctic cities and towns are not only economic or environmental but deeply rooted in sociocultural dynamics that affect the sense of identity, belonging, and empowerment among Arctic populations. In this context, another interviewee from Iceland, said:

> The thing is the people who live in Arctic, you know, in these what you say the periphery, these people want to have better lives, they want to have better houses, they want to have better chances. But then the people who come in from the cities, they tend to think:" Oh, no, you should stay original. You should stay in contact with nature". Tourists expect us to live in line with this romantic idea, which never was true. Of course we are changing constantly, but there is a kind of pressure on that, we should stay the same—what means bad roads and no electricity.
>
> (Researcher, Akureyri, 27 May 2022)

This quote indicates that sustainable development (sometimes wrongly understood, e.g., as a need to avoid modern advanced infrastructure) and expectations from outside the Arctic can exert pressure on cities and municipalities and limit their development in Arctic remote and even non-remote areas.

Power imbalances and the related missing political representation of remote cities also mirrors the gaps in the political representation at the individual level, such as the limited inclusion of perspectives from minorities and, in contrast, the over-representation of individuals, those so-called "local policy entrepreneurs" who are connected beyond their local "bubble" and form part in exclusive networks. Also, in our case study cities in the European Arctic, knowledge exchanges often take place in closed circles (see Chapter 5). To overcome this power imbalance and over-representation of some citizens, better education (1) and social awareness (2) as well as a good coordination and engagement of various actors in contact with the regulatory sector (3) seem crucial. Cases from South Africa and the Philippines, for example, illustrated that people who participate in policy-making do not necessarily understand the trade-offs resulting from political decisions due to a lack of education and information (SUDEA 2022). Also, our interviewees stressed that citizens often apply a local perspective only and do not consider broader consequences. Educational challenges, however, are multifaceted, ranging from gender disparities in educational attainment to the critique of tertiary institutions. These institutions are sometimes perceived as falling short of their social missions, not fully addressing the unique needs of their Arctic contexts or contributing effectively to local development. A politician from Kolari (Politician, Kolari, 7 March 2022) stated, for example: "And I can say from my village when they [representatives from the administration/DW] are in my community when they make this zoning for the mine everything what we said was perhaps read, but immediately forgotten".[8]

Education, however, is also seen as needed to challenge global perspectives that dominate local perspectives. This can be seen in SUD rankings for the Arctic, which often emphasise technological aspects but are not based on local perspectives, as also the concept of smart cities illustrates (SUDEA 2022). However, due to the region's peripherality, scarce infrastructure, and vast distances, our interviewees highlighted that people from the Arctic often have to move for higher education:

> The thing is that [it] is an island and we're quite far from everyone else. We are completely isolated and it has always been very difficult for us to get the international connections. We had to move to another country to get educated. And it was a long trip, and it's only in the last 15 years or so that you can actually get cheap flights. Before that, you would go away and you would come maybe at a Christmas or not even at Christmas. It would be maybe [a] year until you came back.
>
> (Researcher, Akureyri, 27 May 2022)

Irrespective of education, local actors will only get involved in such processes if they (are able to) trust that negotiations are reliable, relevant, and meaningful. Our interviewees and also various speakers at our workshop highlighted that in several cases the related discussions have been impeded by (1) social conflicts and emotions (e.g., in Kolari between local communities, local authorities, and external stakeholders), (2) by national political and economic interests and by international influences (including EU regulation in the European context, e.g., the European Green Deal), (3) by new economic trends like ongoing green energy transformation, and (4) by global developments or geopolitical crises. In view of the latter, surprisingly, the literature on remote cities did not relate much to the effects of the COVID-19 pandemic and the war in Ukraine on interactions among humans but stated only general implications for the SDGs, for example, in view of agriculture and food systems (Adhikari *et al.* 2021). In contrast and as sketched in Chapter 6, the war against Ukraine has changed the tradition of the Arctic as a space of cooperation and affected the long-term perspective for sustainable development in the circumpolar Arctic (including Russian territories) negatively (SUDEA 2022). And also, the pandemic and related travel restrictions enforced a "lock-in" in the cities under investigation as seasonal employees and international tourists did not visit or stay in the places for some time.

In view of (2) and (3), our interviewees also related to postcolonial processes, amongst others to accusations on Green Colonialism in Sápmi. Particularly in conversations about Nuuk and Greenland at large, a recurring focus was on the emerging forms of oppression impacting identity and culture, which perpetuate social divisions. Such situations—as our interviewees pointed out—can sometimes be noticed in the sphere of activities of the public administration and the Danish companies in terms of large infrastructure projects. Such dynamics not only underscore the existing social divisions but also highlight the challenges in ensuring public participation in the planning and execution of SUD strategies.[9] Our interviewees' statements also included various perspectives on environmental sustainability, including comparisons of Iceland to a laboratory for renewable energy and sustainable development and discussions on the sensitivity of Arctic nature to climate change. Moreover, the difficulties of integrating SUD within the unique climate challenges faced by the Arctic regions were raised. One respondent, for example, felt that linking the Arctic to climate change is a challenge or obstacle to the functioning of cities, from which the world demands much more than it demands of itself.

> in the Arctic, you think about the sustainable development agenda as any other city in the world, meaning you relate to the sustainable development goals as the biggest umbrella issue and then translate it into the local context but here the local context is different, sustainable development is different in Arctic cities than in any other city in the world, but we have not managed to analyse and to define the sustainability agenda in the Arctic context because we live in an area where this problem of climate change is pushing us to relate to it" due to the consequences climate change in the Arctic has for other places on Earth.
>
> (City administration, Tromsø, 24 March 2023)

A researcher from Iceland, on the other hand, found that many positive changes are happening in the Arctic and that the region is, in fact, a pioneer when it comes to sustainable development (Researcher, Akureyri, 27 May 2022). As can be seen, there is no consensus among European Arctic city dwellers when it comes to assessing the changes taking place and their impact on SUD in the Arctic.

8.3.2 How to Overcome the Lack of Formalised Structures?

As was also stressed in the literature on remote regions, many interviewees favoured more spaces and political will to transfer and co-produce knowledges, and to integrate diverse perspectives. Both spaces and political will are important to advancing more holistic policy-making across governance levels without neglecting place-sensitivities and power imbalances. But how should such structures look like?

As we have shown in the previous chapters, at the local level, our interviewees emphasised a lack of political will to pursue the global goals and a lack of participation and transparency in policy-making processes. Both seem to require more coordination—by the cities themselves and also by national governments that were seen as hiding away from their responsibilities (e.g., in Kiruna and Kolari): "Sustainability and climate goals are defined by the nation-state. But then the way the municipalities are going to reach these two goals are like in a way up to them" (Researcher, Kiruna, 20 December 2021). At the same time, to overcome the lack of capacity (financial resources and knowledges), the literature and our interviewees see the need to empower local governments to organise meaningful and inclusive participatory processes: often, the engagement of citizens in participatory processes has been considered superficial due to lack of interest, lack of expertise, distrust, critical approaches presented in mass media, and old-fashioned communication channels (SUDEA 2022). The consideration of different phases in participatory processes (in planning and implementation) might be a first important step, as particularly participation in the implementation seems often challenging and requires continuing commitment and resources (SUDEA 2022). It is also important to acknowledge, however, that more inclusive participation in policy-making does not necessarily lead to "more just" policy outcomes. Citizen councils, which promote inclusiveness because participants are randomly selected and the councils "are future oriented and encourage slow thinking", could be a promising format to provide the space needed for inclusive knowledge transfers (SUDEA 2022). However, as long as citizen councils are not integrated into formal decision-making, it will not be clear whether the recommendations developed by citizen councils just shine on paper or have a real impact. For that, legal frameworks would have to change and to define the mandate of citizen councils (SUDEA 2022).

To strengthen knowledge transfer on SUD across governance levels and regions, which is essential for shared understanding, the Nordic model could be adapted at the regional level. Especially under the Nordic model (see Chapter 5), cities are key agents in designing change and despite all differences, there are certain comparable aspects that could be improved (e.g., lack of transnational cooperation, meaningful participation). Also, in view of city-to-city cooperation, cooperation patterns in the Arctic often follow the direction from the South to the North (Southern institutions facilitating North-South cooperation) instead of East–West and thus do not bring political authority/ownership to the region (SUDEA 2022). As shown in the previous Chapters 6 and 7, transnational cooperation between cities in the European Arctic is mainly declarative and in practice underdeveloped, which means that its potential remains almost untapped. Surprisingly, however, from our interviewees, only a few related to economic benefits arising from the level of cities, even though, in the opinion of some interviewees, there is potential for closer cooperation, as exemplified by the various activities carried out in the framework of the West Nordic Cooperation. Many Arctic communities, for example, experience demographic change and have difficulties to attract professionals, which is why incentivising investments from outside are needed, which ideally also helps to diversify the often imbalanced economies of Arctic cities.[10] In a similar vein, our interview data highlights the complexities of economic development, sociocultural challenges, and the unique nature of infrastructure in the Arctic, with additional layers of interest provided by postcolonial and geopolitical themes.

8.4 Conclusions: Transferability of Findings on Remote Regions and Further Research

While the Arctic has often been framed as an "exceptional region", we showed above that in policy-making for SUD many challenges that are visible in the (European) Arctic also apply to other remote regions and vice versa, which is why we argue for that both strands of literature (Arctic studies and studies on peripheral or remote regions) hold a great potential to cross-fertilise one another. A number of similarities emerge, particularly with regard to the two main aspects introduced above: the relational embeddedness of places (1) and the interactions between people that shape and are shaped by the dynamic nature of cities (2).

In view of the relational embeddedness of remote cities (1), often a lack of connectedness among urban places in remote regions is caused by limited (train) transportation, expensive airborne transportation,[11] by insufficient and expensive internet connectivity and problems with access to social and health services. In addition, however, also due to a low population density (e.g., in Lapland, less than two persons per square kilometre) and long distances to capital cities, people feel that they need to find local solutions to challenges,[12] which limits the political will to align policies with those of other regions. This also relates to the limited interactions between people (2) beyond their "bubbles", strong regional identities and limited transnational networks. There is an awareness of external (global, international) expectations,[13] but the responses have more to do with a place branding strategies than actual activities and urban transformations.[14] Limited political representation often seems to go hand in hand with a lack of capacities, which is also caused by limited economic development and dependencies, a lack of diversification of the local economies, insufficient financial resources due to low incomes from taxes, dependency on national government and state public institutions and a lack of qualified people.[15]

All these aspects help to explain the absence of transnational efforts and limited inclusive participatory governance approaches in the European Arctic and in other remote regions. Given the factors introduced in Chapter 3 to advance the pursuit of the global goals, we further found that in light of the institutional set-up and capacities (1), education and (institutional) learning as well as more spaces to advance knowledge transfers are needed. In view of actors (2), it seems that in sparsely populated areas, only few people are able to become "policy entrepreneurs" because of missing connectivity and a lack of access to closed networks or clubs. This missing connection also explains why political priorities (3) more often concern local issues, as citizens feel that they are not included in policy-making processes across governance levels and feel dominated by dependencies they consider unable to change.

Despite all the limitations that come along with the place-sensitivities highlighted in the literature and also in our empirical data, we thus consider our model applicable also to other remote regions. Due to a lack of empirical data, however, we could not provide strong evidence on the relevance that we ascribe to the variable that remoteness remains the basis of imaginaries that firmly define, frame and fuel SUD in the European Arctic and also in other remote regions (key driver 1, Chapter 3). Insights shared during our workshop indicate transferability but further research will be needed to evaluate how this driver shapes SUD in other remote regions specifically. The same applies to key driver 2 (Chapter 3), that cooperation stimulates the pursuit of global goals in remote regions. However, there is a strong indication that this driver also applies to other remote regions, as the limited connectivity and alignment with global goals may have a causal link.

Acknowledgements

This chapter was written by Dorothea Wehrmann. Michał Łuszczuk wrote parts of Section 8.3. The outline was developed by Michał Łuszczuk and Dorothea Wehrmann. Heike Großer (IDOS) supported the literature research for this chapter. The chapter was reviewed by Jacqueline Götze, Katarzyna Radzik-Maruszak, Arne Riedel, and an external expert. All authors of this book have read and accept the content of this chapter.

Notes

1 As explained in Chapter 1, in our research, remoteness relates to places that are characterised by rurality, sparse population, and limited infrastructure. We acknowledge that geographic and economic remoteness often interlink and that also environmental, political, and cultural conditions shape the remoteness of places.

2 Aiming to shed light on the constraints and prospects for sustainable urban development within the European Arctic, our study delved into the widely discussed concept of "Arctic exceptionalism". While the questionnaire did not explicitly inquire about this notion, the broader context of the Arctic regions was recurrently brought up by participants.

3 Code 1: European Arctic and other remote regions (also in Chapter 1). Code 2: Arctic and remote regions.

4 The literature review considered articles listed by SCOPUS, OPAC, and DEBIS until January 2024 when searching for the following keywords in the period 2020–2024: (1) remote regions/remote areas/remoteness/peripheral regions/peripheral areas/isolated regions/remote urban places/cities/municipalities, (2) sustainable development/sustainability/sustainable urban development/urban development, (3) transnational cooperation/city network(s)/knowledge exchange, (4) local governance/participatory approaches/transparency/stakeholder inclusion/trust/knowledge exchange, (5) global goals/climate targets/sustainability goals, (6) multi-level governance/governance across levels/policy alignment across governance levels, (7) SDGs: society/climate targets: society, (8) multi-actor approaches/multi-stakeholder approaches/actor diversity/multi-actor partnerships/multi-stakeholder partnerships, and (9) orchestration/indirect governance.

5 Sustainable local development, in contrast, has been studied since the 1980s (Milán-García *et al.* 2019).

6 For example, in the European Arctic, "the self-perceptions of cities and their means to participate in networks differ a lot" (SUDEA 2022).

7 The object of exploitation in the Arctic is becoming space and physically related resources, including cheap energy, which is used, for example, in aluminum factories in Iceland. The strain on local communities resulting from such practices is not a recent development. As highlighted by an Icelandic interviewee, it emerged earlier than the sustainability discussions, making the main arguments in some public debates primarily workplace wage issues: "We have several of these large aluminum plants and large foreign companies are coming here and they buy cheap energy. They were actually here before we started to talk seriously about global warming or anything in that area. At that time, it was economic thing—more jobs, so we built these really huge aluminum plants and we do the hydroelectric dams and then it's a great combo" (Researcher, Akureyri, 27 May 2022).

8 Interviewees further mentioned that for political reasons the municipality had not organised exchanges that bring together people with different views. As a consequence, such seminars have been organised privately (cf. i.e., Politician, Kolari, 21 November 2022).

9 The Danish-Greenlandic context is also overlaid with postcolonial activities originating, for example, in Iceland (so-called West Nordic postcolonialism). One of the respondents from Akureyri mentioned that while Greenland remains under Denmark's control, such

as the Faroe Islands, Icelanders have been making their presence known in Greenland by distinguishing themselves from the Danes. They have leveraged this presence to seize opportunities in tourism and fisheries, acting in a manner reminiscent of colonial overlords by monopolising these opportunities for themselves.

10 As revealed through our data, the sociocultural landscape of Arctic cities is deeply influenced by demographic trends, educational challenges, and an almost pervasive sense of peripheral neglect. Key demographic processes, such as the migration of populations towards larger urban centres—illustrated by the movement of people, particularly those with higher education levels, from northern locales to southern regions in countries like Iceland and Norway underscore the centrifugal forces reshaping the Arctic's demographic profile. Interestingly, a slight reversal of this trend was noted in Finland, indicating a nuanced demographic dynamic across the European Arctic region.

11 "It is very hard to get out from the North; if you live in smaller places, the plane tickets are very expensive (as to go to London); but there is no alternative for flights" (Politician, Tromsö, 20 March 2023).

12 "[O]ne aspect of how space (remoteness) impacts on local governance is 'smallness' (of state) in terms of space and population; when smallness goes with remoteness than you have a problem, because you can't really ask someone else for help because you're isolated. So, you have to take care of it yourself" (Researcher, Akureyri, 27 May 2022).

13 "(…) people are starting to understand too much bigger extent how important the Arctic context is and that this is a new era for us in kind of I would say empowerment. We're trying to engage ourselves much more in our Arctic future in our framework conditions, we're understanding much more that if we want to have a say and if we want to have ownership and responsibility for our own lives, our own futures then we do need to understand that our existence is not just local, regional, or national. It is international and it is pan Arctic. And it has to do with international law with United Nation. It has to do with Brussels framework conditions in Brussels and the EA agreement. It has to do with influencing national capitals. But it also has to do with trying to get more of the conversation about the Arctic future to the Arctic" (Administration, Tromsö, 24 March 2023).

14 For example, the new Finnish strategy emphasises Finland as an Arctic state. The Arctic is presented as being of great relevance not only to Finnish Lapland but also to places in the South of Finland (Politician, Rovaniemi, 15 February 2022).

15 "[…] you only have a limited pool of people to recruit a full government and regional governments and all of that. I think that also Icelandic people are sometimes hard pressed to find really qualified personnel for everything that they need if they want to provide full services that the country needs to provide. So occasionally they get quite unqualified governments" (Researcher, Akureyri, 27 May 2022).

References

Adhikari, J., Timsina, J., Khadka, S. R., Ghale, Y., and Ojha, H., 2021. COVID-19 impacts on agriculture and food systems in Nepal: Implications for SDGs. *Agricultural Systems*, 186, 102990. https://doi.org/10.1016/j.agsy.2020.102990

Cavalcante, A. M., Marquezini, M. V., Mendes, L., and Moreno, C. S., 2021. 5G for remote areas: Challenges, opportunities and business modeling for Brazil. *IEEE Access*, 9, 10829–10843. https://doi.org/10.1109/ACCESS.2021.3050742

Coenen, L., Hansen, T., Glasmeier, A., and Hassink, R., 2021. Regional foundations of energy transitions. *Cambridge Journal of Regions, Economy and Society*, 14 (2), 219–233. https://doi.org/10.1093/cjres/rsab010

Frank, K. I., and Hibbard, M., 2017. Rural planning in the twenty-first century: Context-appropriate practices in a connected world. *Journal of Planning Education and Research*, 37 (3), 299–308. https://doi.org/10.1177/0739456x16655599

Fu, H., *et al.*, 2022. The spatial pattern and governance of Zhongyuan Urban-Rural system in its development trajectory. *Journal of Geographical Sciences*, 32 (7), 1261–1280. https://doi.org/10.1007/s11442-022-1996-3

Götze, J., 2024. *Sámi-EU relations: Sámi participation in EU policy-making as an example of Indigenous peoples' participation in transnational policy-making*. University of Bonn. https://doi.org/10.48565/bonndoc-362

Groth, S., Klinger, T., and Otsuka, N., 2023. Geographies of new mobility services: The emergence of a premium mobility network space. *Geoforum*, 144, 103765. https://doi.org/10.1016/j.geoforum.2023.103765

Hao, J., *et al.*, 2020. Investigating the accessibility between civil airports and tourist locations in tourist cities in Yunnan Province, China. *Sustainability*, 12 (10), 3963. Available from: https://www.mdpi.com/2071-1050/12/10/3963

Harvey-Scholes, C., Mitchell, C., Britton, J., and Lowes, R., 2023. Citizen policy entrepreneurship in UK local government climate emergency declarations. *Review of Policy Research*, 40 (6), 950–971. https://doi.org/10.1111/ropr.12522

Hawken, S., Rahmat, H., Sepasgozar, S. M. E., and Zhang, K., 2021. The SDGs, ecosystem services and cities: A Network analysis of current research innovation for implementing urban sustainability. *Sustainability*, 13 (24), 14057. Available from: https://www.mdpi.com/2071-1050/13/24/14057

Kosai, S., and Yamasue, E., 2021. Towards intercity cooperation: Comparison of spatial transport energy efficiency between central and peripheral cities in Japan. *In:* Kishita, Y., Matsumoto, M., Inoue, M., and Fukushige, S., eds., *EcoDesign and sustainability II: Social perspectives and sustainability assessment*. Singapore: Springer Singapore, 239–253.

Leane, E., *et al.*, 2021. From gateway to custodian city: Understanding urban residents' sense of connectedness to Antarctica. *Geographical Research*, 59 (4), 522–536. https://doi.org/10.1111/1745-5871.12490

Lindberg, M., *et al.*, 2020, Co-creative place innovation in an arctic town. *Journal of Place Management and Development*, 13 (4), 447–463.

Marschall, P., Wingens, C., and Dick, E., 2021. *Kommunale Entwicklungspolitik in Deutschland (Vol. Studies* 105*)*. Bonn: German Development Institute / Deutsches Institut für Entwicklungspolitik (DIE).

Milán-García, J., Uribe-Toril, J., Ruiz-Real, J. L., and de Pablo Valenciano, J., 2019. Sustainable local development: An overview of the state of knowledge. *Resources*, 8 (1), 31. Available from: https://www.mdpi.com/2079-9276/8/1/31

Patel, Z., 2022. The potential and pitfalls of co-producing urban knowledge: Rethinking spaces of engagement. *Methodological Innovations*, 15 (3), 374–386. https://doi.org/10.1177/20597991221129779

Raufirad, V., Hunter, R., Khalili, R., and Bagheri, S., 2017. Drivers of local people's participation in sustainable natural resource management: A case study in central Iran. *Local Environment*, 22 (7), 880–893. https://doi.org/10.1080/13549839.2017.1297391

Scholvin, S., 2021. Peripheral development through extractive industries? Gateways, local content policy, and the oil and gas sector in Neuquén and Río Negro, Argentina. *Applied Geography*, 135, 102554. https://doi.org/10.1016/j.apgeog.2021.102554

Smith, D. A., and Barros, J., 2021. Chapter 3—Sustainable transport planning and residential segregation at the city scale. *In:* Mulley, C., and Nelson, J. D., eds., *Urban form and accessibility*. Oxford: Elsevier, 27–44.

SUDEA. 2022. *Advancing Sustainable Urban Development in Remote Regions*. Summary report. 25 May 2022. Available from: https://www.idos-research.de/fileadmin/user_upload/pdfs/veranstaltungen/2022/20220525_SUDEA_workshop_report.pdf [24 November 2024].

Vale, M., *et al.*, 2023. Are peripheral regions in troubled waters for sustainability transitions? A systematic analysis of the literature. *European Urban and Regional* Studies, 31(2), 09697764231194316. https://doi.org/10.1177/09697764231194316

Wang, H.-J., 2023. Smart city branding vision: Multiple stakeholder perspectives. *Innovation: The European Journal of Social Science Research*, 1–25. https://doi.org/10.1080/13511610.2023.2296384

Wang, Y., Zhao, P., Xie, S., and Zhang, W., 2023. Mesoscale structure in urban–rural mobility networks in the Pearl River Delta area: A weighted stochastic block modeling analysis. *ISPRS International Journal of Geo-Information*, 12 (5), 183. Available from: https://www.mdpi.com/2220-9964/12/5/183

Xiao, H., Bao, S., Ren, J., and Xu, Z., 2023. Transboundary impacts on SDG progress across Chinese cities: A spatial econometric analysis. *Sustainable Cities and Society*, 92, 104496. https://doi.org/10.1016/j.scs.2023.104496

Yu, H., *et al.*, 2022. Morphological and functional polycentric structure assessment of megacity: An integrated approach with spatial distribution and interaction. *Sustainable Cities and Society*, 80, 103800. https://doi.org/10.1016/j.scs.2022.103800

Zeigermann, U., Kammerer, M., and Böcher, M., 2023. What drives local communities to engage in climate change mitigation activities? Examining the rural–urban divide. *Review of Policy Research*, 40 (6), 894–919. https://doi.org/10.1111/ropr.12528

Zhou, Y., *et al.*, 2023. The mechanism behind urban population growth and shrinkage from the perspective of urban network externalities. *Chinese Geographical Science*, 33 (2), 189–204. https://doi.org/10.1007/s11769-023-1340–6

9 Conclusions

9.1 Introduction

This chapter primarily summarises and discusses the main findings and outcomes of the four-year German-Polish interdisciplinary SUDEA project that began in the autumn of 2019. The project aimed to identify and explain the reasons behind the misalignment of sustainable urban development policies across governance levels in the European Arctic. This research problem was connected with two assumptions. We presumed that the remoteness of the cities located in the European Arctic is a fundamental obstacle for their engagement in pursuing global agendas. At the same time, we suspected that long-established (transnational) collaborative frameworks in the circumpolar Arctic support the interpretation of the global goals at the national and local levels. This project focused on seven cities located in the European Arctic—Rovaniemi, Kolari, Nuuk, Akureyri, Tromsø, Kiruna, and Luleå—and explored how dynamics in multilevel governance, specifically multi-actor cooperation and participation in policy-making processes, shape local approaches to sustainable urban development in remote regions (see Chapter 1).[1]

Building on the analysis presented in the previous chapters, we integrate the main findings and synthesise key insights here. Our goal is to contribute to a more nuanced understanding of how sustainable urban development is pursued in remote regions and how policy-making may be advanced to ensure that global goals are met while addressing local needs and priorities. Additionally, we provide several policy recommendations and ideas for further research about the complexities of sustainable urban development in remote regions. Generally, we highlight the misalignment between global visions shaped by the Paris Agreement, the New Urban Agenda, and the 2030 Agenda for Sustainable Development and local realities and imaginaries that—as our analyses show—are characterised by power imbalances and weak participatory approaches. Furthermore, we investigate the diverse capacities and specific contexts in which actors operate and emphasise the need for more inclusive and participatory governance approaches.

This chapter begins by summarising how our interviewees described local approaches to sustainable urban development in the European Arctic. We discuss this description of the status quo in light of the theoretical perspectives introduced in Chapters 2 and 3, highlighting the importance of inclusive and participatory

DOI: 10.4324/9781003284864-9

governance and vibrant regional collaboration, where integrating various actor perspectives is vital for creating sustainable urban environments. We then critically inspect the conceptual model proposed in this book, pointing out its transferability, limitations, and gaps. This sets the stage for subsequent policy recommendations and suggestions for future research directions.

Structurally, this chapter is divided into five key sections. After this introduction, we revisit the main research questions and assumptions laid out in Chapter 1, providing a context for the detailed analysis that follows. The core of this part is dedicated to the findings of the SUDEA project, where we explore and explain the significant discrepancies between global sustainable development goals and local realities; this includes insights into specific barriers and enablers of policy implementation. We continue with a presentation of policy recommendations designed to enhance sustainable development through better alignment of local and global policies. We underscore the need for improved transnational cooperation, strengthened local capacities and more inclusive participatory processes. We conclude this chapter with an overview of a list that includes several topics and ideas we discovered during our research that we could not investigate in detail because they went beyond the scope of our project. Nevertheless, they are all suggestions for future research directions and, hopefully, inspiration for other researchers fascinated with the Arctic regions and their challenges.

This chapter not only provides a critical evaluation of the current situation and the existing political and governance frameworks but also proposes actionable recommendations to bridge the gap between global aspirations and local implementations, aiming to foster more sustainable and inclusive urban development in the European Arctic and similar remote regions. We are convinced that the findings from our research are not only relevant to the European Arctic but also offer valuable lessons for other remote areas facing similar challenges. In the next section, we present the key findings from our research, highlighting the critical issues and insights that have emerged from our analysis.

9.2 Local Approaches to Sustainable Urban Development in the European Arctic: Insights Revealed by SUDEA

Sustainable urban development in the European Arctic, particularly in the face of significant challenges posed by climate change, presents a complex but crucial endeavour intricately tied to the region's unique historical, socio-economic, and environmental contexts. It involves balancing environmental preservation with social well-being and economic growth. It requires innovative approaches to land and water management, infrastructure development, energy use, transport, public services, and, last but not least, community engagement. The mentioned aspects, along with many others not detailed here, must also consider the diverse values, needs, and interests of various actors in the cities investigated and also at other governance levels, such as national governments and the international community. By envisioning sustainability, all these actors navigate the complexities of development in the world's most vulnerable and dynamic regions. A critical question arises here: To what extent is this navigation influenced—facilitated or hindered—by global arrangements that establish the principles, directions and goals intended to support sustainable development on a worldwide scale? In line with other studies, our findings from the European Arctic illustrate that sustainable urban development policies are not aligned across governance levels, which means that the efforts undertaken by various actors are far from synergistic, weakening the pursuit of the global goals. As mentioned, our research—focused on seven cities located in the European Arctic (Rovaniemi, Kolari, Nuuk, Akureyri, Tromsø, Kiruna, and Luleå)—has revolved around three overarching questions. The detailed justification for this research, the assumptions made and the description of the project are presented in Chapter 1. At this point, however, it is necessary to recall the main research points and questions and provide the answers we sought.

It is important to recall here that our research was anchored in three global instruments, which, according to common political debates as well as the academic literature, aim to shape a specific framework determining the scope, direction, and scale of sustainable urban development globally. The Paris Agreement, the New Urban Agenda, and the 2030 Agenda have nearly worldwide coverage and tackle policy- and decision-making across various levels of governance; concurrently, their legal nature and respective implementation methods of their respective goals differ significantly (details are presented in Chapter 4).

In light of our research problem, the pursuit and implementation of the global goals at the local level is of fundamental importance. We have explained this issue in detail in Chapter 2 pointing out explicitly how the three instruments relate to governance across multiple levels and suggest an inclusive, participatory multi-actor approach to achieve their envisioned implementation. The objectives and proposed methods of implementation emphasised by the 2030 Agenda, the Paris Agreement, and the New Urban Agenda are all formed and grounded in a multilevel governance perspective. Moreover, cooperation coordination for enhancing action and adaptation is considered vital for policy alignment and bringing different actors together.

All three instruments also emphasise an inclusive and participatory approach, which should serve as the foundation for the agreed-upon goals and their envisioned implementation. In this context, we argue that both the Paris Agreement and the 2030 Agenda address the country level in particular, but they do not highlight differences within countries, among rural and urban, and among central and remote areas, which are of great relevance when considering the existing diversity. In contrast, the New Urban Agenda is unique in underlining "the key role of cities and human settlements as drivers of sustainable development in an increasingly urbanized world" (UN 2017, p. 11).

At the same time, we stress that all three instruments adopt a context-sensitive approach, consider the varying capacities of the actors involved in implementing the global objectives and promote capacity development and capacity building. In the context of sustainable urban development policy, this approach potentially offers adequate flexibility and sensitivity to the differing circumstances of local policies. However, it also shifts responsibility to national governments and local authorities to meet expectations or targets that occasionally appear unrealistic in their situations. It generates the risk that applied solutions may fall short of expectations, leading to more discouragement than motivation for further activity. As we show in Chapter 2, the New Urban Agenda is the only instrument taking a local and pluralistic stance to actor engagement when promoting "capacity development as a multifaceted approach that addresses the ability of multiple stakeholders and institutions at all levels of governance" (UN 2017, p. 37).

During our field research, we observed that the perception of international arrangements, including of the three instruments, varies significantly at the local level. We have grouped these differences into four topic categories. Firstly, it is how global visions are integrated into national legislation and communicated in/with municipalities and to what extent adherence to sustainability requirements can be encouraged at the national level, as well as in economic terms. Secondly, local priorities often conflict with agendas at other governance levels. Thirdly, it is self-perception; small, remote Arctic urban communities frequently feel that they cannot significantly contribute to necessary developments because they are primarily affected by global changes, which are caused outside the region. Fourthly, there is a limited expectation that sustainable urban development recognition can enhance visibility and create a positive "brand" for remote cities, aligning local ambitions with global expectations.

Based on our analysis, we argue that global sustainable development instruments provide an essential conceptual and practical framework for shaping sustainable urban development in the European Arctic. However, our examination reveals apparent gaps between the aspirations and the realities of implementing these provisions. These gaps were also highlighted in numerous interviews, which often indicated that sustainable urban development is not perceived as a responsibility for all citizens. Additionally, there was criticism that global perspectives and ambitions do not align with local conditions and are not always necessary on a local scale, where proximity to the environment is more immediate. While these assessments are generalisations, our research has identified several key differences (and similarities) crucial for understanding the incoherence between local sustainable urban development policies and global goals.

While the issue of the limited or mainly declarative presence of global agreements at the level of local policies is a conclusion from the analysis of documents and interviews, the results regarding the reasons why this happens are much more nuanced. This brings us to the following question—*why do approaches to sustainable urban development in the selected cities differ?* Our conceptual model and the qualitative analysis of official documents from the cities and interviews conducted provide the following answers:

First of all, governance in all cities investigated is shaped by the Nordic model of governance and its principles of actor engagement. As explained in Chapter 4, the model exemplifies a balanced approach where local autonomy, democratic engagement, and welfare provision are expected to be harmoniously integrated. The engagement of different actors is claimed to be a cornerstone of this model, ensuring that governance is inclusive, responsive, and reflective of the community's needs. The collaborative and transparent nature of this model is expected to foster trust and accountability, making it a robust framework for local governance, also in the field of sustainable urban development. Bearing in mind that the model refers to specific values, rules, and general institutional arrangements, during our field research, we noticed differences between the cities in how these solutions are implemented. These differences clearly aligned with our conceptual model, as they were primarily concerned with institutional set-up and capacities including (1) intergovernmental relations, (2) policy capacities, (3) institutional structure within a city, (4) citizens' participation and political representation on a local level, (5) financial, infrastructural, and technological resources, and (6) participation in city networks and partnerships. Fewer discrepancies could be seen in terms of the actors involved and the roles they performed, for example, generally strong position of economic actors or low involvement of indigenous peoples' representatives. Perhaps the most similar factor turned out to be political priorities in the field of sustainable development. The reason for this could be the reference to the global instruments, the use of the exact keywords, as well as the high level of generality of these declared priorities or goals. Alike were also the critical voices of city residents regarding—in general—the slow progress in advancing urban sustainable development programmes.

Furthermore, field research made it possible to explore the relevance of imaginaries in shaping sustainable urban development policies (the first of the two key drivers presented in our model), which determined the approach of all actors to the issue of sustainable urban development, climate change, and the directions of socio-economic development of cities. These imaginaries, on the one hand, allow citizens to notice the changes taking place nearby. But, on the other hand, they are also often arguments for ignoring the challenges related to climate change, accessing energy resources, and, finally, respecting the rights of Indigenous peoples, just to mention a few main issues.

In turn, as highlighted in Chapter 5, peripherality in connection with limited institutional and human capacities proved to be a barrier for scrutinised cities for full engagement in (transnational) cooperation, which, in our model, is another driver supporting better alignment between local policies and global goals and ambitions.

The results of our research, which turn out to be a considerable challenge in the context of the functioning of the Nordic model in the studied cities, are observations related to local participation. All municipalities in our case study cities implement participatory mechanisms to varying extents. However, local approaches to urban development do not seem to build on engaged local participation. This is due to power relations, language barriers, differences in the level and type of knowledge(s), as well as competencies related to discussing technical aspects, cultural communication styles, and common low interests in the public sphere policy-making. As it was shown in Chapter 2, "ordinary citizens" are among the weakest actors, both in the context of knowledge and available resources, which makes their engagement very limited. This, in turn, makes the participatory processes not perceived to be meaningful, which is why many citizens do not see how they could actually influence them. Our research shows that the smaller the city in terms of population and socio-economic potential, the greater the doubt about the significance of such involvement. Among the analysed cities, only Luleå, Rovaniemi, and Tromsø stand out for their active and innovative approaches due to their larger size, economic activities, and international contacts, and also for being academic and tourist hubs.

To sum up, the approaches to sustainable development in selected cities diverge due to different relationships occurring within the three interlinked factors in policy-making and decision-making processes (actors and their relationships, institutions and their set-ups, and political priorities). These relations are deeply embedded in imaginaries, and their dynamics and scope are determined by cities' specificity and limited resources. Remoteness is another element that keeps these cities out of the mainstream by making sustainable urban development inferior in importance to more urgent and current issues. Although remote cities try to meet the expectations of national authorities at the level of political declarations, even local politicians and administration representatives admit that the possibilities of their actions are inadequate to the expressed plans and intentions. Local approaches do not prioritise/follow solutions consistent with the goals formulated above, which simultaneously leads to the establishment of local solutions that are unique and potentially challenging to transfer to other locations.

This brings us to this book's next key research question, namely, *how can sustainable Arctic urban development be pursued in alignment with local and global policies?*

As we explained in Chapter 2, the 2030 Agenda, the Paris Agreement, and the New Urban Agenda prioritise two fundamental governance approaches: governance across multiple levels and an inclusive, participatory multi-actor approach, despite their differences. These approaches are integral to the framing of the global goals and are deemed crucial for their effective implementation. Multi-level governance is critical for providing access to finances, knowledge(s), social, and basic services while promoting coordination, integration, and participation across governance levels. An inclusive and participatory approach, the foundation of these global goals, is vital for their successful realisation. As we show in Chapter 2, the three instruments stress the importance of cooperation across different levels of governance, highlighting the necessity of strengthening institutional frameworks. They specifically emphasise the critical roles of national, subnational, and local governments and their partnerships with other public bodies and non-governmental organisations. The emphasis is on fostering coordination and cooperation to facilitate engagement and partnerships at all levels. This cooperation aims at developing effective policies that are context-sensitive, at recognising different capabilities, and at upholding the principle of shared responsibility. Capacity building is a critical component in pursuing these goals. By reinforcing institutional arrangements and encouraging participation, these global governance instruments seek to create a more integrated and effective approach to addressing global challenges

Furthermore, cooperation is widely understood as a complex political and academic construct that operates across various governance levels (see Chapter 2). It is necessary to pursue the global goals, as challenges need to be addressed by various actors across different governance levels and regions, as highlighted by SDG 17. In the European Arctic, however, cooperation to pursue the global goals is still limited among different actors and across regions; specifically, its local implementation tends to default to passive policies and "business as usual" due to administrative limitations, the challenges of managing urban areas in harsh and remote climates, and difficulties in attracting skilled personnel. This gap between high-level policy and local implementation hampers effective cooperation. Particularly transnational cooperation often seems unimportant at the local level—a sentiment also echoed in the literature on city networks. This perception may arise from a lack of visible, immediate benefits or understanding of such cooperation's value to local communities. However, there is a contrast between the local immediacy offered by networks like the Arctic Mayors' Forum (AMF) and the broader perspective provided by entities like the Arctic Council. While the former is expected to deliver immediate answers and a sense of closeness, the latter provides a more global environmental and political outlook. Balancing these perspectives and integrating them into coherent policies and actions is challenging.

As discussed in previous chapters, Arctic administrative bodies often drive cooperation, which can lead to discontinuity and a lack of consistent focus on the unique challenges of northern peripheries. Frequent changes in officials and local politicians, along with their shifting priorities and strategies, exacerbate this issue. This underscores the need for greater political engagement, a deeper understanding of the local context and more direct citizen participation in decision-making. These issues highlight the structural, operational, and perceptual challenges facing transnational cooperation in the Arctic. In Chapters 2 and 6, we argue that overcoming these obstacles requires a concerted effort to strengthen local capacity, enhance operational efficiency in cooperative forums, reconcile local and global perspectives, and encourage sustained and informed political leadership.

From our perspective, more specifically we propose to (1) enhance context-sensitive and inclusive participation to avoid dichotomous discussions and feelings of misunderstanding or social exclusion. Increased national support for municipal decision-making is essential to address current deficits in participatory practices, ensuring that no one is left behind. This involves addressing the overrepresentation of economic priorities, improving the design of participatory processes, and balancing the involvement of different actors. We suggest (2) strengthening local capacity, delegating funds and more monitoring from the national to the local level. Such action can foster greater local independence and ownership of sustainable urban development initiatives. This approach would address the imbalance of tasks and resources, reduce external domination, and ensure a more balanced representation of priorities and actors. Finally, we propose (3) enhancing Arctic regional cooperation and learning can complement national governance efforts regarding funding and capacity building. Such collaboration can support municipal capacities in participatory practices while promoting local self-determination by reducing dependencies on national support. Encouraging regional cooperation could be facilitated by funding schemes addressing the design of participatory processes, the imbalance of tasks and resources, and external domination.

Building on the previous discussions and characterisations of the factors and key drivers, we propose in Chapter 6 three potential pathways for enhancing sustainable development by linking vertical and horizontal alignment. These pathways adopt a multilevel governance approach, particularly emphasising polycentric governance, which incorporates specialisation, division of tasks and subsidiarity while addressing local-regional circumstances and community preferences to create more efficient policies. Achieving this requires access to finances, knowledge, social, and basic services, which necessitates coordination, integration, and participation. The analysis of our data, based on the conceptual model developed in Chapter 2, identified capacity building and leadership as crucial for developing policies aimed at enhanced sustainable (urban) development.

Finally, we explored *to what degree approaches to sustainable urban development in the European Arctic are specific or typical for remote regions more generally.*

While the Arctic has often been considered an "exceptional region", our analysis demonstrates that the challenges seen in sustainable urban development in the (European) Arctic are also present in other remote regions and vice versa. This suggests that Arctic studies and studies on peripheral or remote regions have great potential to cross-fertilise each other. Several similarities emerge, particularly concerning two main aspects: the relational embeddedness of places (1) and the interactions between people that shape and are shaped by the dynamic nature of cities (2).

Considering the former, the lack of connectedness among urban places in remote regions is often due to limited (train) transportation, expensive airborne transportation, insufficient and costly internet connectivity, and problems with accessing social and health services. Additionally, low population density and long distances to capital cities necessitate local solutions to challenges, limiting the political will to align policies with other regions. This isolation also relates to limited interactions between people beyond their immediate circles, strong regional identities, and limited transnational networks. While there is awareness of external (global, international) expectations, responses often focus more on place branding strategies than on actual urban transformations. Considering the latter, limited political representation frequently coincides with a lack of capacities, which is further exacerbated by limited economic development and diversification, insufficient financial resources due to low tax incomes, dependence on the national government and state public institutions, and a shortage of qualified people. These factors explain the lack of transnational efforts and inclusive, participatory governance approaches in the European Arctic and other remote regions.

In light of the factors introduced in Chapter 2, advancing global goals requires addressing institutional set-ups and capacities (1) by improving education and (institutional) learning and creating more spaces for knowledge transfer. Regarding actors (2), in sparsely populated areas, only a few individuals can become "policy entrepreneurs" due to limited connectivity and access to closed networks. This disconnection also explains why political priorities (3) often focus on local issues, as citizens feel excluded from policy-making processes across governance levels and dominated by dependencies they believe they cannot change.

Despite these limitations and place-specific sensitivities, we believe our model can be applied to other remote regions, too (see Chapter 7). However, due to a lack of empirical data, we cannot provide strong evidence on the relevance of remoteness as a foundational aspect that defines sustainable urban development in the European Arctic and other remote regions. Insights shared during our experts' workshop indicate the model's transferability, but further research is needed to evaluate how this variable specifically shapes sustainable urban development in other remote regions.

Similarly, while key driver 2 suggests that cooperation stimulates the pursuit of global goals in remote regions, more research is needed to confirm this variable's applicability elsewhere. However, there is a strong indication that limited connectivity and alignment with global goals may also have a causal link in other remote regions.

9.3 Evaluating the Conceptual Model "Local Approaches to Sustainable Urban Development in Remote Regions": Advantages and Limitations

In this section, we evaluate the conceptual model for sustainable urban development in remote regions, highlighting its advantages and limitations. Developing this model was challenging, as it required synthesising extensive literature research and empirical findings. At the same time, the model turned out to be useful and demonstrated the operational ability to identify key drivers and factors in the analysis of local approaches to sustainable urban development. It functions both when comparing individual remote cities and in relation to the four key dimensions of the studied phenomena, namely environmental, social, economic, and geopolitical dimensions. A synthetic summary of the results of such operationalisation of the model is provided in Table 9.1.

Despite its advantages, the model also has certain limitations. Initially, we acknowledge that we were aware of some of the shortcomings, deficiencies, and risks associated with such a tool. These include the problems demonstrated by Ong and Jabbari (2019), among which are inconsistency between different components of a model or between multiple models, hidden dependencies between elements of a model, and high complexity affecting ambiguity and data integration. Moreover, the authors argue that "[a] single model cannot fully suffice system representations as they often only represent either the structure, behaviour, or executory aspects" (Ong and Jabbari 2019, p. 12). Moreover, we were also aware of the concerns standing behind the question about

> the role that the researcher exerts in the face of theoretical models: should the researcher be an ambassador for a given model, or rather act as an independent examiner who challenges the model with his own ideas and methods, checks up its workings and strives to better understand behavioural processes?.
>
> (André and Laurencelle 2020, p. 1)

Table 9.1 Operationalisation of the SUDEA-model in the context of four dimensions.

	Key dimensions of studied developments			
	Environment	*Economy*	*Society*	*Geopolitics*
Key trends observed during the project lifetime	Rapidly changing landscape (to different extends at different places)	Economic transformations	Demographic change	Heightened security tensions after February 2022
Driver 1 Imaginaries (=internal driver of change): envisioned future development of the city and respective interventions needed	Environmental transformations affect the use of space and its landscape (new challenges—where to store snow in compact cities, how to continue reindeer herding near the city infrastructure or how to keep sidewalks free of ice if temperatures are constantly around 0°C?)	Infrastructure will adapt according to economic focus (tourism, mining, wind power plants), will change the environmental landscape, will affect the job market, housing situation and provision of social services	Empty space: outward migration: young people move South, ageing society, services cannot be obtained. Growing space: inward migration: newcomers from the South/other places, more housing and services needed	Growing fears in the areas close to border with Russia; growing militarisation of territory
Driver 2 Transnational cooperation and global arrangements (=external driver of change): access to knowledge, practice and experiences, to resources (human and financial, reputation and reporting)	Growing aim to identify the common challenges and risk, mitigation and adaptation measures	Cities and national governments use different narratives to frame the issues at stake	Advancement of societal relations across the borders is driven by individual engagement	Efforts to lower the tensions and provide support for people in need
Factor 1 Actors and their relationships. Who is in charge? What are competences, capacities, interests? How are decision-makers embedded in networks?	Example Tromsø: Discussion on public transportation and toll system. Green party in favour but new party formed just to fight the planned toll system	Example Kolari: A city council composed with a majority of former miners favours mining instead of tourism.	Example Nuuk: Informal links and private channels of communication	Growing pressure from military, industry, and energy sectors

(*Continued*)

Table 9.1 (Continued)

	Key dimensions of studied developments			
	Environment	*Economy*	*Society*	*Geopolitics*
Factor 2 Institutions and their set-ups. How flexible or static are institutional set-ups? How are (short-, mid-, and long-term) processes organised?	Tromsö, Akureyri, and Rovaniemi are university-cities and "capitals" of the North	Example Kiruna: Economic and political domination of mining company; Example Akureyri: lack of strong industry	Everywhere: limited presence and visibility of NGOs	Lack of well-developed mechanisms for negotiating with the military forces
Factor 3 Political priorities. Do incentive structures, budget decisions favour one dimension more than the other?	Need to develop urban planning to meet the environmental implications for the economy and society, including traditional livelihood.	Resolving the dilemma: economic transformation in harmony or at the expense of social preferences or the environment (Arctic Development Paradox)?	Including the human and non-human perspectives in spatial development plans	Dealing with the current dominance of the security narrative and the risks of disinformation; preliminary exploration of ways to restore pan-Arctic cooperation

Source: Own work.

Basing our model on the results of previous research and enriching it with our findings, we were therefore aware of some of the emerging limitations. Through iterative activities on the construction of the model and discussions with other researchers, we sought to optimally construct the model and, at the same time, learn about its weaknesses.[1] In what follows, we first briefly discuss a few limitations that can be identified in the model and second assess the applicability of the model beyond the (European) Arctic.

The main concerns regarding the model can be organised into three strands. Firstly, we observe two conceptual shortcomings. The first is the underemphasis on the role of the national government, which plays an essential part in setting the legal, political, and financial framework for (sustainable) urban development at the local level. Moreover, national governments are also gatekeepers of sorts in terms of how cities participate in cooperative formats (see more: Wehrmann *et al.* 2022b). While addressing the impacts of the national-level in Chapter 5 and including them in the model as a part present in all three factors, for example, when discussing the inter-governmental relations (see point 2.2), the national component could be more visible in the model. Another gap is related to the participation of Indigenous peoples' representatives or organisations in policy-making processes of sustainable urban development. Indigenous peoples' role in the Arctic's multilevel governance is a fundamental issue and has been explored in numerous studies. However, our field research did not recognise any channels that would especially empower the Indigenous voices in sustainable urban development in the European Arctic municipalities (Mattsson and Götze 2022, Nystø Keskitalo and Götze 2023). Therefore, the lack of this element is not a research oversight but a consequence of the current state of affairs, which—in our opinion—should also be improved in this respect. Just as sustainability in the Arctic should not be imagined without considering the knowledges, interests, and rights of Indigenous peoples, their influence on urban development should not be limited either. Currently, there are situations where, due to factors such as specialised or technical vocabulary, the participation of representatives of indigenous peoples is hampered.

Secondly, we are aware that the model is partially insufficient in exploring the non-political dimension of local approaches to sustainable urban development. While we highlighted the importance of geographical and environmental imaginaries, we only briefly mentioned questions of identity, social and cultural values, economic factors, and transnational economic exchanges. This shortcoming is a consequence of the specificity of our research, and we can only support doing new research about these aspects.

Thirdly, it is not entirely clear to what extent the impact of climate change may influence the model. On the one hand, it is clear that the currently observed consequences of climate change affect the way cities in the (European) Arctic operate, and thus, they can determine the approaches and political priorities formulated in sustainable urban development policies. On the other hand, our research reveals that, in reality, there is no apparent connection between the effects of climate change and the level or scope of activity through participatory mechanisms or in (transnational) cooperation. It is undoubtedly another limitation of our model and, at the same time, a significant research problem.

Finally, assessing the model's applicability beyond the Arctic, which was discussed in Chapter 7, we argue that the model might help research SUD in remote regions where power imbalances and limited political representation impede policy alignment across governance levels and areas. The model recognises that remote cities are affected by a lack of capacities (financial resources and knowledge), which partly explains why participatory processes can be considered superficial. To become "portals of knowledge exchanges", local governments worldwide must be empowered to organise meaningful and inclusive participatory processes, provide space for knowledge transfers and anchor institutional learning. Moreover, the literature review suggests that it is prevalent that local priorities often seem detached from policy priorities at other governance levels. The plurality and different social realities are not reflected in sustainable urban development processes designed beyond the local level, which have adverse effects on the political will to pursue such priorities. Variations in local approaches to SUD seem encouraged more by individual "local policy entrepreneurs" than by "local civic capacity", which is often limited due to sparse infrastructure, weak connectivity and missing access to higher education.

We are also convinced that the model's applicability beyond the (European) Arctic is supported by including imaginaries with vast potential for including different local features and contexts. However, it can be claimed that remote resource-rich areas worldwide are often portrayed as a target of external exploitation and resource extraction. Such perceptions reinforce a sentiment of domination and provoke social objections.

Additionally, for citizens in remote regions around the world, it seems more challenging to move beyond their "local bubble" due to power imbalances (centre-periphery but also among actor groups) and limited capacities (e.g., to maintain and build up inward and outward relationships). The research and studies on other remote regions provide little evidence on formalised structures that facilitate cooperation even though coordinated network approaches are widely seen as beneficial for sustainable development. Interactions among actors at different places seem limited, but some are better connected than others ("gateways"). It is also possible to note a "clubbing" effect among remote cities that are more strongly connected.

9.4 Policy Recommendations for Applying the Model in the Context of Sustainable Urban Development in the European Arctic

This section presents a series of policy recommendations derived from an extensive analysis of sustainable urban development in remote regions, with a particular focus on the European Arctic presented in this book. The insights gained from this research primarily reflect the unique and common challenges faced by the seven cities examined; to a lesser degree, they are also informed by insights into broader contexts of other remote areas globally. By synthesising data from case studies, interviews, and literature reviews, we have identified key areas where policy interventions might significantly enhance sustainable urban development outcomes, bring it closer to people, and better align with global instruments. These observations are closely linked to the three pathways proposed in Chapter 6.

The primary objective of these recommendations is to address the critical gaps and power imbalances that impede effective governance and sustainable urban planning in remote regions. To be more precise, these recommendations aim to foster more inclusive, participatory approaches to sustainable urban development. Through these proposals, we emphasise the importance of enhancing connectivity of the remote regions (both in national and transnational dimensions), supporting local governance capacities, and promoting socio-economic diversification. By doing so, we hope to contribute to creating more sustainable urban models that can be adapted to various remote contexts beyond just the Arctic.

We would like to highlight that the significance of these policy recommendations extends beyond immediate practical implications. They are intended to serve as inspiration for the development of strategic frameworks for policy-makers, urban planners, and local governments to advance sustainable urban development collaboratively. We believe that by implementing these policies, different actors can ensure that remote regions are not left behind in the global push towards sustainability and climate change adaptation and mitigation. Instead, these areas can become exemplars of how tailored, context-sensitive policies can drive substantial progress in achieving the SDGs.

Ultimately, our proposals put forth in this section aim to provide actionable and evidence-based guidance to enhance the quality of life in remote regions. They are designed to empower local communities, improve multilevel governance structures accordingly with democratic values and create robust frameworks for long-term sustainable development. Through these efforts, we aspire to contribute to a more equitable and sustainable future for remote and peripheral regions worldwide.

Based on our research and discussion during the workshop in Bonn (May 2022) (Wehrmann *et al.* 2022a) and Bodø (May 2024), we propose the following policy recommendations focusing on strengthening the effectiveness and inclusivity of sustainable urban development policies at various governance levels. These recommendations are linked to the pathways introduced in Chapter 7 and are ordered according to their level of detail starting with the most general and then moving on to more specific ones. Implementing most of these recommendations requires financial resources, so the first step is to allocate appropriate funding for the proposed activities.

9.4.1 Strengthening Cooperation

There is a constant need to foster stronger cooperation between national, regional, and local governments to better align sustainable urban development strategies. This involves creating formal intergovernmental and transnational dialogue structures (like special councils, committees, joint working groups, or regular conferences or expert meetings), sharing best practices, and ensuring that local governments have the resources and support needed to implement national and international sustainability commitments. As we argue in Chapter 2, the most promising solutions can be found in the orchestration concept, as "the platform logic broadens the view of network governance to a broader set of connections, the orchestration of multiple logics and ecosystem thinking" (Haveri and Anttiroiko 2023, p. 17) and supports the idea of direct or participatory democracy (Sahamies *et al.* 2022). This recommendation is particularly relevant to Pathway 1: Nordic Model 2.0 and Pathway 3: Strengthen the Nexus of Regional Cooperation & the Design of Global Agreements.

9.4.2 Develop Resilient Multilevel Governance Structures

We propose developing multilevel governance structures that are more resilient and adaptable to external shocks, such as geopolitical tensions, economic collapse, or pandemics. This involves establishing flexible policies and protocols that can quickly respond to changing circumstances and ensure continuity in cooperation efforts. Of course, we know that, under the principle of subsidiarity, the most appropriate structures for this are those located as close as possible to the levels where the problems occur. At the same time, our research clearly shows that mechanisms also of a supranational nature are needed to address cross-border challenges. For many issues related to the consequences of climate change, beneficial solutions may appear in closer regional cooperation, which will be approached in the following recommendations. This proposal aligns with Pathway 1: Nordic Model 2.0 and Pathway 2: Towards More Inclusive and Active Multi-Actor Platforms.

9.4.3 Establish Localised Indicators and Monitoring Systems

We suggest establishing localised indicators and monitoring systems for the SDGs to bridge the gap between national indicators and local implementation. This would enable municipalities to track progress more accurately and tailor actions to their specific contexts. Moreover, such indicators could facilitate cooperation between cities, as they would allow the identification of places and related good practices worth popularising. Again, such a monitoring system could be embedded in transnational institutions or programmes. Some first similar initiatives in that directions have been implemented by Nordregio and its partners (Nordregio 2024). This recommendation supports Pathway 1: Nordic Model 2.0 and Pathway 3: Strengthen the Nexus of Regional Cooperation & the Design of Global Agreements.

9.4.4 Develop More Robust Cooperation Frameworks

We recommend the development of more robust frameworks for cooperation between Arctic cities, focusing on structured and continuous collaboration to enhance policy coherence and address shared challenges in sustainable urban development. This aligns with Pathway 2: Towards More Inclusive and Active Multi-Actor Platforms. Appreciating the creation of the AMF, we still see a lot of room for its development, both in the subjective and objective, internal and external dimensions. It is worth making efforts to ensure that the organisation unites many members, allowing not only for the intensification of horizontal connections among them but also for the strengthening of the AMF's voice in the vertical system, that is, towards the states of the region and the Arctic Council. Realising that the strength of such transnational initiatives lies primarily in the quality of the contribution formulated by and in individual municipalities, we point out possibilities of strengthening the local dimension of sustainable urban development policies in subsequent recommendations.

9.4.5 *Provide Targeted Training for Local Officials*

Even in the Nordic countries, there is a persistent need to provide targeted training and resources for local government officials to enhance their capacity to implement sustainable urban development initiatives. This includes training on new planning tools, sustainable practices, and ways to engage with communities and stakeholders effectively, and last but not least, to engage in transnational cooperation. This recommendation is crucial for Pathway 1: Nordic Model 2.0 and Pathway 3: Strengthen the Nexus of Regional Cooperation & the Design of Global Agreements. Moreover, we claim that increasing resources and supporting local governments is indispensable for effectively managing and implementing participatory processes, ensuring they have the necessary skills and knowledge. At the same time, we propose encouraging local policy entrepreneurs and civic leaders to take active roles in shaping sustainable development policies, ensuring that local needs and perspectives are adequately represented.

9.4.6 *Advance Genuine Inclusive and Participatory Governance*

We propose advancing genuine inclusive and participatory local governance and decision-making processes dedicated to sustainable urban development. This can be achieved through, for example, enhancing public participation in urban planning processes and development processes. This includes mandatory public consultations organised at a time available for many people, innovative participatory tools and platforms, accessible dissemination of adequate (easy to understand) information in all appropriate languages, and opportunities for co-development with community members to ensure that planning reflects the needs and preferences of local populations. Moreover, we propose semi-formalised participatory structures, such as citizen councils, to ensure inclusive and meaningful participation in policy-making processes. These councils should be integrated into formal decision-making frameworks to ensure that their recommendations have a real impact. In this context, we must highlight that all actor groups, including marginalised communities, should have equal opportunities to participate in governance processes. Furthermore, we propose to advance transparency in decision-making by (1) establishing transparent and open communication channels between local governments and citizens, and (2) by mechanisms for holding decision-makers accountable. All these ideas are pertinent to all three pathways—Pathway 1: Nordic

Model 2.0, Pathway 2: Towards More Inclusive and Active Multi-Actor Platforms, and Pathway 3: Strengthen the Nexus of Regional Cooperation & the Design of Global Agreements.

9.4.7 Facilitate Cross-Sector Collaboration

We advocate for facilitating cross-sector collaboration and the integration of diverse knowledge systems, including Indigenous knowledges and scientific expertise, into urban planning and policy-making processes. It can be succeeded by promoting collaboration between public, private, and civil society sectors to leverage diverse perspectives and resources. This can be achieved through inclusive forums like "third spaces" or "portals for knowledge exchange", workshops, and collaborative projects. These structures or programmes should promote continuous dialogue and collaboration among local actors, national policy-makers, transnational partners, and international experts. This recommendation supports Pathway 2: Towards More Inclusive and Active Multi-Actor Platforms and Pathway 3: Strengthen the Nexus of Regional Cooperation & the Design of Global Agreements.

The policy recommendations outlined above aim to advance sustainable urban development in remote regions by addressing governance gaps, fostering inclusivity, enhancing local capacities, and promoting resilient, participatory communities. By linking these recommendations to the pathways introduced in Chapter 6, this approach, hopefully, provides practical, evidence-based strategies for state and non-state actors to drive sustainable progress in the European Arctic and beyond. These strategies should be continually reviewed and developed based on both the monitoring of changing situations and needs, as well as inspired by the conclusions of ongoing and future scientific research (Berrone *et al.* 2023). With this in mind, the next section offers suggestions for researchers interested in further exploring the issues considered in this book.

9.5 Future Research Avenues

This section outlines a series of ideas and proposals for further research derived from our comprehensive analysis of sustainable urban development in the European Arctic as a remote region. The goal is to identify critical knowledge gaps, guide scholars in their efforts to deepen the understanding of sustainable urban development in remote regions, and propose areas where future research can provide valuable insights into the ongoing and expected scholarly debates to progress in sustainable urban development and cooperation.

By highlighting specific concepts and topics that require further investigation, we aim to encourage targeted research that can address the unique challenges faced by these regions. These recommendations emphasise inter alia the importance of understanding the impact of connectivity, the role of local policy entrepreneurs, the effectiveness of participatory governance structures, and the strategies for economic diversification. By focusing on the most pressing and under-researched issues, we aim to foster a collaborative approach to research that includes local communities, governments, and international organisations. This collaborative research can lead to more effective and context-sensitive solutions, ultimately contributing to the sustainable development goals in remote regions.

Ultimately, these ideas for further research are addressed, not just to an academic audience but also to a broad readership including policy-makers, urban planners, and development practitioners, such as those in NGOs. By prioritising these research areas, we hope to stimulate a deeper investigation into the complex dynamics of sustainable urban development in remote regions. This will not only advance academic knowledge but also support the creation of more resilient, inclusive, and sustainable urban environments in some of the world's most challenging and unique regions. Below, we outline several crucial and impactful topics for further research.

9.5.1 Intergovernmental and Transnational Cooperation

There is a need for critical analysis of the current level and perspectives of further intergovernmental and transnational cooperation for sustainable development. While our findings support earlier research (Moallemi *et al.* 2020, Abbott and Snidal 2021, Biermann *et al.* 2022, Hölscher and Frantzeskaki 2020), we see a necessity to deepen knowledge about dynamics, hindrances, and good practices in implementing sustainable development policies and focusing on the interplay between national, regional, and local governments.

9.5.2 Local Government Involvement in Climate Change Adaptation

It is important to better explore the involvement of local governments in climate change adaptation and mitigation, focusing on their specific roles and contributions in achieving national climate change mitigation targets; this includes case studies of successful and unsuccessful local initiatives that have already been undertaken (Henfrey *et al.* 2023, Goodwin *et al.* 2023, Sareen and Waagsaether 2023, Masuda *et al.* 2022).

9.5.3 Effectiveness of Participatory Governance Structures

There is still a knowledge gap about the effectiveness of participatory governance structures. We propose investigating the efficacy of different participatory governance models, such as citizen councils, in enhancing inclusivity and accountability in sustainable urban development decision-making processes (Newig *et al.* 2023, Cloutier *et al.* 2015, Hügel and Davies 2020, Cattino and Reckien 2021, Wu *et al.* 2020). Additionally, we suggest examining (1) how informal governance practices, such as decisions made in informal settings, influence formal decision-making processes and (2) the representation and participation of marginalised groups, including Indigenous peoples, immigrants, and seasonal workers, in local governance structures and decision-making processes. Additionally, there is a need to scrutinise and evaluate the effectiveness of various participatory tools and methodologies, such as participatory budgeting and urban laboratories, in promoting sustainable urban development in (European) Arctic cities.

9.5.4 Integration of Indigenous Knowledges

Future research should be more dedicated to exploring the integration of Indigenous knowledge systems into urban planning processes in (European) Arctic cities and their impact on sustainable development outcomes (Berkes *et al.* 2007, Vogel and Bullock 2021, Kohsaka and Rogel 2021, Vlasova *et al.* 2021, Berg-Nordlie *et al.* 2022). Additionally, the contributions of local policy entrepreneurs in promoting sustainable development initiatives in remote regions, including their strategies and challenges, should be examined (Capano and Galanti 2021, Esteves *et al.* 2021, Veleva 2021). Finally, the influence of the different industries present in the circumpolar areas (mining, data science, tourism, etc.) on urban development policies and participatory processes in Arctic municipalities can be explored more often (Hall 2020, WoodDonnelly and Ohlsson 2023).

9.5.5 Impact of Public Participation

There is a need for thorough studies and comparisons on the effectiveness of various public participation mechanisms in influencing urban development outcomes, particularly regarding sustainability, inclusivity, and resiliency (Wu *et al.* 2020, Åström 2020, Hofer and Kaufmann 2023, Sandin 2020).

9.5.6 Economic Impacts of Sustainable Urban Development

We observe an increasing necessity to explore and better understand the economic impacts of sustainable urban development initiatives. This includes analysing the costs and benefits of implementing green infrastructure, renewable energy projects and sustainable transportation systems (Natcher and Koivurova 2021, Wood-Donnelly and Ohlsson 2023, Garbis *et al.* 2023, Filippova *et al.* 2022, Mortensen *et al.* 2017).

9.5.7 Multi-Actor Cooperation Dynamics

Studying the dynamics of multi-actor cooperation and knowledge exchange between remote regions is crucial for achieving sustainable development goals. Hence, research should focus on the best practices and barriers to effective collaboration, including geopolitical tensions such as the Russian war in Ukraine (Steinveg *et al.* 2024, Guo *et al.* 2024, Andreeva and Rottem 2024, Dyck 2024, Łuszczuk *et al.* 2023, Hilde *et al.* 2024).

9.5.8 Effectiveness of City Networks

Examining the effectiveness of city networks like the AMF in achieving sustainable urban development goals and facilitating knowledge exchange among Arctic cities is vital (Wehrmann *et al.* 2022b, Leffel *et al.* 2023, Filimonova 2024) to advancing shared understandings and political leverage.

9.5.9 Impact of Connectivity Improvements

We also highlight a need to investigate how developments in transportation and digital connectivity influence sustainable urban development outcomes in remote regions, focusing on the economic, social, and environmental impacts (Natcher and Koivurova 2021, Haugland *et al.* 2023, Cho and Jull 2023).

These proposed research ideas are designed to build a comprehensive understanding of the unique challenges and opportunities in remote regions, particularly in the European Arctic. By addressing these topics, future research can provide valuable insights and practical solutions that support the sustainable urban development agenda. These efforts are crucial for ensuring that remote and peripheral regions are not only included in but also contribute significantly to the sustainable development goals and are able to provide good living conditions for their citizens.

Researchers, policy-makers, and practitioners are encouraged to collaborate across disciplines and borders to develop innovative approaches that address these critical areas. This collaborative effort will help create resilient, inclusive, and sustainable urban environments, ensuring that these remote regions can thrive in the face of global challenges. We believe that by fostering such an integrated approach, we can better support the development of tailored policies and practices that meet the specific needs of remote regions, ultimately contributing to a more equitable and sustainable future. With these research avenues identified, we can now conclude by summarising the broader implications of our study and reinforcing the importance of continued collaboration and innovation in this field.

9.6 Conclusions

This chapter underscored the critical importance of aligning global sustainable development goals with local realities in the European Arctic. The findings from the SUDEA project revealed significant gaps between the aspirations of global instruments and the practical challenges faced by remote Arctic municipalities. Despite the unique socio-economic, environmental, and cultural contexts of these regions, there is a clear need for more robust and inclusive governance structures that can bridge these gaps and foster sustainable urban development.

This chapter provided a nuanced understanding of the complexities of achieving sustainable development in remote regions by highlighting the specific issues of power imbalances, limited political representation, and the fact that local priorities often conflict with agendas at other governance levels. The insights gained from this research not only enabled us to formulate policy recommendations but also to emphasise the need for further research to explore innovative solutions and best practices. It is imperative that local, national, and international non-state and state actors collaborate more effectively, leveraging diverse perspectives and resources to create resilient, inclusive and sustainable urban environments.

Ultimately, the path to sustainable urban development in the European Arctic and other remote regions lies in integrating local knowledge(s) and global objectives supported by robust participatory processes and transnational cooperation. By addressing these challenges head-on, policy-makers, researchers, and practitioners can contribute to a more sustainable and equitable future, ensuring that remote regions are not left behind in the global pursuit of sustainability and climate resilience. This book, therefore, serves as a call to action, urging all actors to commit to the shared goal of sustainable urban development through informed, inclusive, and collaborative efforts.

Acknowledgements

This chapter was written by Michał Łuszczuk. The chapter was reviewed by Jacqueline Götze, Dorothea Wehrmann, and two external experts. All authors of this book have read and accepted the content of this chapter.

Note

1 We are particularly grateful for critical comments from the participants of the Arctic Congress 2023, who participated in the session organised by Prof. A. Petrov under the title "Sustainability Research and Practice in the Arctic: Concepts, Approaches, Opportunities and Emerging Agenda" (Bodø, Norway, 30/05/2024).

References

Abbott, K.W., and Snidal, D., 2021. Strengthening international regulation through transnational new governance. *In:* Abbott, K. W., and Snidal, D., eds., *The spectrum of international institutions. An interdisciplinary collaboration.* Abingdon: Routledge, 95–139.

André, N., and Laurencelle, L., 2020. Where do the conceptual models for behaviour change come from, and how are they used? A critical and constructive appraisal. *The Quantitative Methods for Psychology*, 16 (1), 1–8. DOI: 10.20982/tqmp.16.1.p001.

Andreeva, S., and Rottem, S. V., 2024. How and why the Arctic Council survived until now – An analysis of the transition in chairship between Russia and Norway. *The Polar Journal*, 14 (1), 229–246. DOI: 10.1080/2154896X.2024.2342111.

Åström, J., 2020. Participatory urban planning: What would make planners trust the citizens? *UP* 5 (2), 84–93. DOI: 10.17645/up.v5i2.3021.

Berg-Nordlie, M., Dankertsen, A., and Winsvold, M., eds., 2022. *An urban future for Sàpmi? Indigenous urbanization in the Nordic states and Russia.* New York: Berghahn Books (Studies in the circumpolar North, 4).

Berkes, F., Berkes, M. K., and Fast, H., 2007. Collaborative integrated management in Canada's north: The role of local and traditional knowledge and community-based monitoring. *Coastal Management*, 35 (1), 143–162. DOI: 10.1080/08920750600970487.

Berrone, P., *et al.*, 2023. How can research contribute to the implementation of sustainable development goals? An interpretive review of SDG literature in management. *International Journal of* Management Reviews, 25 (2), 318–339. DOI: 10.1111/ijmr.12331.

Biermann, F., Hickmann, T., and Sénit, C.-A., eds., 2022. *The political impact of the sustainable development goals.* New York: Cambridge University Press.

Capano, G., and Galanti, M. T., 2021. From policy entrepreneurs to policy entrepreneurship: Actors and actions in public policy innovation. *Policy & Politics,* 49 (3), 321–342. DOI: 10.1332/030557320X15906842137162.

Cattino, M., and Reckien, D., 2021. Does public participation lead to more ambitious and transformative local climate change planning? *Current Opinion in Environmental Sustainability*, 52, 100–110. DOI: 10.1016/j.cosust.2021.08.004.

Cho, L., and Jull, M., eds., 2023. *Design and the built environment of the Arctic*. London: Routledge.

Cloutier, G., *et al.*, 2015. Planning adaptation based on local actors' knowledge and participation: A climate governance experiment. *Climate Policy*, 15 (4), 458–474. DOI: 10.1080/14693062.2014.937388.

Dyck, C., 2024. On thin ice: The Arctic Council's uncertain future. *Marine Policy*, 163, 106060. DOI: 10.1016/j.marpol.2024.106060.

Esteves, A. M., *et al.*, 2021. Sustainable entrepreneurship and the sustainable development goals: Community-led initiatives, the social solidarity economy and commons ecologies. *Business Strategy and the Environment*, 30 (3), 1423–1435. DOI: 10.1002/bse.2706.

Filimonova, N., 2024. Constructing climate change: Exploring how cities frame climate change in the Arctic. *Journal of Urban Affairs*, 1, 1–19. DOI: 10.1080/07352166.2024.2351391.

Filippova, N. A., Vlasov, V. M., and Bogumil, V. N., 2022. Transport planning and sustainable development in the Arctic region. *In:* Pak, E. V., Krivtsov, A. I., Zagrebelnaya, N. S., eds., *The handbook of the Arctic*. Singapore: Springer Nature Singapore, 833–843.

Garbis, Z., *et al.*, 2023. Governing the green economy in the Arctic. *Climatic Change*, 176 (4), 1–23. DOI: 10.1007/s10584-023-03506-3.

Goodwin, S., Olazabal, M., Castro, A. J., and Pascual, U., 2023. Global mapping of urban nature-based solutions for climate change adaptation. *Nature Sustainability*, 6 (4), 458–469. DOI: 10.1038/s41893-022-01036-x.

Guo, Y., Bai, R., and Hong, T., 2024. Transboundary cooperation in Arctic climate change governance under geopolitical tensions. *Journal of Environmental Management*, 358, 120855. DOI: 10.1016/j.jenvman.2024.120855.

Hall, H. M., 2020. Innovation, new technologies, and the future of the circumpolar North. In*:* Coates, K. S., and Holroyd, C., eds., *The Palgrave handbook of Arctic policy and politics*. Cham: Springer International Publishing, 117–132.

Haugland, B. T., Ryghaug, M., and Søraa, R. A., 2023. Framing intelligent transport systems in the Arctic: Reindeer, fish and the engineered road. *Engineering Studies*, 15 (1), 50–70. DOI: 10.1080/19378629.2023.2169612.

Haveri, A., and Anttiroiko, A.-V., 2023. Urban platforms as a mode of governance. *International Review of Administrative Sciences*, 89 (1), 3–20. DOI: 10.1177/00208523211005855.

Henfrey, T., *et al.*, 2023. Rethinking the sustainable development goals: Learning with and from community-led initiatives. *Sustainable Development,* 31 (1), 211–222. DOI: 10.1002/sd.2384.

Hilde, P. S., Ohnishi, F., and Petersson, M., 2024. Cold winds in the north: Three perspectives on the impact of Russia's war in Ukraine on security and international relations in the Arctic. *Polar Science*, 41, 101050. DOI: 10.1016/j.polar.2024.101050.

Hofer, K., and Kaufmann, D., 2023. Actors, arenas and aims: A conceptual framework for public participation. *Planning Theory*, 22 (4), 357–379. DOI: 10.1177/14730952221139587.

Hölscher, K., and Frantzeskaki, N., 2020. *Transformative climate governance*. Cham: Springer International Publishing.

Hügel, S., and Davies, A. R., 2020. Public participation, engagement, and climate change adaptation: A review of the research literature. *WIREs Climate Change*, 11 (4), e645. DOI: 10.1002/wcc.645.

Kohsaka, R., and Rogel, M., 2021. Traditional and local knowledge for sustainable development: Empowering the Indigenous and local communities of the world. *In:* Filho, W. L., *et al.*, eds., *Partnerships for the goals*. Cham: Springer International Publishing (Encyclopedia of the UN Sustainable Development Goals), 1261–1273.

Leffel, B., Derudder, B., Acuto, M., and van der Heijden, J., 2023. Not so polycentric: The stratified structure & national drivers of transnational municipal networks. *Cities*, 143, 104597. DOI: 10.1016/j.cities.2023.104597.

Łuszczuk, M., Rakowski, P., and Szkarłat, M., 2023. Poland's polar policy in the face of the Arctic Council 2022 cooperation crisis. *Polish Polar Research*, 44 (4), 339–364. DOI: 10.24425/ppr.2023.146739.

Masuda, H., Kawakubo, S., Okitasari, M., and Morita, K., 2022. Exploring the role of local governments as intermediaries to facilitate partnerships for the sustainable development goals. *Sustainable Cities and Society*, 82, 103883. DOI: 10.1016/j.scs.2022.103883.

Mattsson, M., and Götze, J., 2022. How to develop inclusive, sustainable urban spaces in the European Arctic and beyond – Insights from Kiruna. Available from: https://www.thearcticinstitute.org/develop-inclusive-sustainable-urban-spaces-european-arctic-beyond-insights-kiruna/ [updated on 1/22/2024, accessed 6/11/2024].

Moallemi, E. A., *et al.*, 2020. Achieving the sustainable development goals requires transdisciplinary innovation at the local scale. *One Earth*, 3 (3), 300–313. DOI: 10.1016/j.oneear.2020.08.006.

Mortensen, L., Hansen, A. M., and Shestakov, A., 2017. How three key factors are driving and challenging implementation of renewable energy systems in remote Arctic communities. *Polar Geography*, 40 (3), 163–185. DOI: 10.1080/1088937X.2017.1329758.

Natcher, D. C., and Koivurova, T., 2021. *Renewable economies in the Arctic*. London: Routledge.

Newig, J., Jager, N. W., Challies, E., and Kochskämper, E., 2023. Does stakeholder participation improve environmental governance? Evidence from a meta-analysis of 305 case studies. *Global Environmental Change*, 82, 102705. DOI: 10.1016/j.gloenvcha.2023.102705.

Nordregio 2024. Agenda 2030 at the local level. Available from: https://nordregioprojects.org/agenda2030local/ [updated on 3/13/2024, accessed 6/11/2024].

Nystø Keskitalo, A. M., and Götze, J., 2023. How to streamline Sámi rights into policy-making in the European Union? Available from: https://www.thearcticinstitute.org/how-streamline-sami-rights-into-policy-making-european-union/ [updated on 2/26/2024, accessed 6/11/2024].

Ong, D., and Jabbari, M., 2019. A review of problems and challenges of using multiple conceptual models. *In:* vom Brocke, J., Gregor, S., Müller, O., eds., *27th European conference on information systems. ECIS 2019*. Sweden: Stockholm-Uppsala, 1–18.

Sahamies, K., Haveri, A., and Anttiroiko, A.-V., 2022. Local governance platforms: Roles and relations of city governments, citizens, and businesses. *Administration & Society,* 54 (9), 1710–1735. DOI: 10.1177/00953997211072531.

Sandin, G., 2020. Lack of participatory effort: On the ethics of communicating urban planning. *UP*, 5 (4), 227–237. DOI: 10.17645/up.v5i4.3445.

Sareen, S., and Waagsaether, K. L., 2023. New municipalism and the governance of urban transitions to sustainability. *Urban Studies*, 60 (11), 2271–2289. DOI: 10.1177/00420980221114968.

Steinveg, B., Rottem, S. V., and Andreeva, S., 2024. Soft institutions in Arctic governance—Who does what? *Polar Record*, 60, 1–7. DOI: 10.1017/S0032247423000335.

UN, New Urban Agenda, 2017. Nairobi: United Nations. Available from: https://habitat3.org/wp-content/uploads/NUA-English.pdf [17 May 2024].

Veleva, V., 2021. The role of entrepreneurs in advancing sustainable lifestyles: Challenges, impacts, and future opportunities. *Journal of Cleaner Production*, 283, 124658. DOI: 10.1016/j.jclepro.2020.124658.

Vlasova, T., Petrov, A. N., and Volkov, S., 2021. Rethinking sustainability monitoring in the Arctic by linking resilience and sustainable development in socially-oriented observations: A perspective. *Sustainability*, 13 (1), 177. DOI: 10.3390/su13010177.

Vogel, B., and Bullock, R. C. L., 2021. Institutions, indigenous peoples, and climate change adaptation in the Canadian Arctic. *GeoJournal,* 86 (6), 2555–2572. DOI: 10.1007/s10708-020-10212-5.

Wehrmann, D., *et al.* 2022a. Advancing sustainable urban development in remote regions. Hybrid expert workshop. 25 May 2022, Uniclub Bonn/Zoom Summary Report. Available from: https://www.idos-research.de/fileadmin/user_upload/pdfs/veranstaltungen/2022/20220525_SUDEA_workshop_report.pdf.

Wehrmann, D., *et al.*, 2022b: Transnational cities alliances and their role in policy-making in sustainable urban development in the European Arctic. *In:* Sellheim, N., Menezes, D. R., eds., *Non-state actors in the Arctic region.* Cham: Springer International Publishing (Springer Polar Sciences), 113–131.

Wood-Donnelly, C., and Ohlsson, J., eds., 2023. *Arctic justice. Environment, society and governance*. Bristol: Bristol University Press.

Wu, L., *et al.*, 2020. Improvement of regional environmental quality: Government environmental governance and public participation. *The Science of the Total Environment,* 717, 137265. DOI: 10.1016/j.scitotenv.2020.137265.

Index

For Product Safety Concerns and Information please contact our EU representative GPSR@taylorandfrancis.com
Taylor & Francis Verlag GmbH, Kaufingerstraße 24, 80331 München, Germany

www.ingramcontent.com/pod-product-compliance
Lightning Source LLC
LaVergne TN
LVHW021123300926
849160LV00010B/838
* 9 7 8 1 0 3 2 2 5 7 5 4 9 *